ACHIEVE

EXAM SUCCESS ■ 5th EDITION

A Concise Study Guide for the Busy Project Manager

DIANE ALTWIES, PMP
JANICE PRESTON, PMP

J.ROSS
PUBLISHING

Direct all inquiries to J. Ross Publishing, 300 S. Pine Island Road, Suite #305,
Plantation, FL 33324. Phone 954-727-9333; Fax 561-892-0700;
Web www.jrosspub.com.

Attention: Corporations, Professional Organizations, Universities, and Colleges.
Quanity discounts are available on bulk purchases of this book. For information
contact: salesandmarketing @jrosspub.com.

TABLE OF CONTENTS

FOREWORD

Professional. Experienced. Knowledgeable. Ethical. Value-adding. Continuous learner. Results-driven. Recognized leader. These are just a few words to describe a Project Management Professional (PMP)®.

Congratulations on your decision to become a PMP. The marketplace recognizes the value of a project manager who is a PMP. Project management job postings routinely state "PMP required." PMI salary surveys show that average project management salaries are higher for PMPs than for non-PMPs.

Now that you've made the decision to become a PMP, the hard part of your journey begins. Professional experience alone is insufficient to pass the PMP exam. Passing the PMP exam requires hours of study to become familiar with *A Guide to the Project Management Body of Knowledge* (*PMBOK® Guide*) and other relevant project management literature. This study guide, *Achieve PMP® Exam Success*, will become an invaluable tool in your exam preparation journey.

The words used above to describe PMPs describe the *Achieve PMP® Exam Success* authors. Diane Buckley-Altwies and Janice Preston together have over 30 years of experience helping project managers just like you pass the PMP exam. They have prepared this excellent study guide using techniques proven to help students pass the PMP exam. It's the tool I wish I had when preparing for the PMP exam. It's the tool I use when I teach PMP exam preparation classes.

Achieve PMP® Exam Success follows the *PMBOK® Guide*'s organizational structure and synthesizes knowledge from that volume, *The Standard for Project Management of a Project* (in *Annex A-1*) contained within it, and related project management literature. This study guide's pages contain concise explanations of each knowledge area, process group, and process. The explanations are easy to read and will help you apply the *PMBOK® Guide* to real-life project management situations. Key vocabulary and terms that you must know for the PMP exam are emphasized. The study guide also identifies and demonstrates formulas that you will need to memorize. All combined, the explanations, vocabulary definitions, and formula demonstrations will help you gain an understanding of how to use tools and techniques to produce process outputs. More importantly, this study guide will help you understand why each output is important and how it impacts other processes.

Simply reading *Achieve PMP® Exam Success* gets you started in your exam preparation. To maximize your studying, review the same *PMBOK® Guide* chapter and the *Achieve PMP® Exam Success* chapter simultaneously. Take the time to complete the learning exercises contained in the study guide. The learning exercises use simplified, real-life examples to illustrate key concepts found on the PMP exam. Carefully compare your answers to the provided answers. Understand why the answer addresses the exercise. Avoid the temptation of simply thinking through each exercise briefly and then looking at the answer.

The *Achieve PMP® Exam Success* authors know that the most important part of your PMP exam preparation is for you to spend hours practicing question answering via a hands-on test-taking experience. They have taken great care in writing each question and thoroughly vetted each answer with PMPs. The questions included both within the book and in the electronic resources mirror those you will encounter on the exam. Devote the majority of your study time to answering the sample questions. When you miss a question, review the carefully written explanations of why one answer is the best answer.

Achieve PMP® Exam Success, when used in conjunction with the *PMBOK® Guide*, contains all the study material required to prepare for and pass the PMP exam. Now it's up to you. Invest the time and effort required to take advantage of this powerful exam preparation tool. I hope to see your copy sitting on your desk dog-eared and worn at the end of your PMP exam preparation. Best wishes for a rewarding and successful PMP journey!

Kristine A. Hayes Munson, MBA, PMP, CIA
Financial Services/IT Project, Program and Portfolio Manager
PMP Exam Preparation Instructor
PMI and PMI-Orange County Volunteer

PREFACE

PMI updates the *PMBOK® Guide* every four years. The most recent update, released in January, 2013, is officially called the *PMBOK® Guide*, Fifth Edition. In June 2015, PMI performed a role delineation study and updated the *PMP Examination Content Outline*. These changes were incorporated into the PMP Exam starting January 2016. We have updated this book to ensure the material provided aligns with the most current PMI Standard and supporting documentation. The PMP exam consists of 200 questions. Twenty-five of those are field testing questions to see if they are appropriate. The 25 trial questions will NOT count towards the pass/fail determination. So only 175 of the 200 questions count towards your score.

The *PMBOK® Guide*, Fifth Edition has been significantly improved from its prior editions. To see a full list of changes from the *PMBOK® Guide*, Fourth Edition, reference *Appendix X-1* of the *PMBOK® Guide*, Fifth Edition.

One of the major changes made in this edition of the *PMBOK® Guide* is the addition of a 10th knowledge area: stakeholder management. The table below highlights the process name changes by each knowledge area, including the new knowledge area.

In comparing the *PMBOK® Guide*, Fourth Edition to the *PMBOK® Guide*, Fifth Edition, you will also note that a few processes have been removed, added, or combined, however, the major structure of the *PMBOK® Guide* processes has stayed in-tact. A total of 47 processes exist in the *PMBOK® Guide*, Fifth Edition, increased from the previous 42 processes in the prior edition. The five process groups have not changed.

Knowledge Area	*PMBOK® Guide* Fourth Edition Processes	*PMBOK® Guide* Fifth Edition Processes
Integration	• Develop Project Charter • Develop Project Management Plan • Direct and Manage Project Execution • Monitor and Control Project Work • Perform Integrated Change Control • Close Project or Phase	• Develop Project Charter • Develop Project Management Plan • Direct and Manage Project Work • Monitor and Control Project Work • Perform Integrated Change Control • Close Project or Phase
Scope	• Collect Requirements • Define Scope • Create WBS • Verify Scope • Control Scope	• **Plan Scope Management** (added) • Collect Requirements • Define Scope • Create WBS • Validate Scope • Control Scope

Knowledge Area	PMBOK® Guide Fourth Edition Processes	PMBOK® Guide Fifth Edition Processes
Time	• Define Activities • Sequence Activities • Estimate Activity Resources • Estimate Activity Durations • Develop Schedule • Control Schedule	• **Plan Schedule Management** (added) • Define Activities • Sequence Activities • Estimate Activity Resources • Estimate Activity Durations • Develop Schedule • Control Schedule
Cost	• Estimate Costs • Determine Budget • Control Costs	• **Plan Cost Management** (added) • Estimate Costs • Determine Budget • Control Costs
Quality	• Plan Quality • Perform Quality Assurance • Perform Quality Control	• Plan Quality Management • Perform Quality Assurance • Control Quality
Human Resources	• Develop Human Resource Plan • Acquire Project Team • Develop Project Team • Manage Project Team	• Plan Human Resource Management • Acquire Project Team • Develop Project Team • Manage Project Team

Communications	• **Identify Stakeholders**
	  (moved)
	• Plan Communications
	• Distribute Information
	• **Manage Stakeholder Expectations**
	  (moved)
	• Report Performance
		• Plan Communications Management
		• Manage Communications
		• Control Communications

Risk	• Plan Risk Management • Identify Risks • Perform Qualitative Risk Analysis • Perform Quantitative Risk Analysis • Plan Risk Responses • Monitor and Control Risks	• Plan Risk Management • Identify Risks • Perform Qualitative Risk Analysis • Perform Quantitative Risk Analysis • Plan Risk Responses • Control Risks

Knowledge Area	PMBOK® Guide Fourth Edition Processes	PMBOK® Guide Fifth Edition Processes
Procurement	• Plan Procurements • Conduct Procurements • Administer Procurements • Close Procurements	• Plan Procurement Management • Conduct Procurements • Control Procurements • Close Procurements
Stakeholder Management	Was not defined	• **Identify Stakeholders** (moved) • **Plan Stakeholder Management** (added) • **Manage Stakeholder Engagement** (moved) • **Control Stakeholder Engagement** (added)

The new knowledge area, stakeholder management, is in Chapter 13 of the *PMBOK® Guide*. Much of the content for this chapter was previously included in the communications management knowledge area. PMI's goal with this new knowledge area is to highlight the importance of effectively engaging all stakeholders, especially those external to the project, in project decisions and execution.

Several general changes have been added throughout the new edition. PMI has developed some basic rules for defining and using inputs, tools, techniques, and outputs (ITTOs) as well as project documents and the project management plan. There is more consistency in the titles of the various processes. Processes in the Monitoring and Controlling process group typically all start with the word Control, and each knowledge area has a planning process that requires the project manager to plan the management of that process group.

Finally, this edition has been harmonized with other PMI standards such as the *Glossary of Terms* with the *PMI Lexicon of Project Management Terms*, which is the central body of terms that all of PMI standards reference.

Now lets look at each chapter individually for some specific changes.

Chapter 2, Project Management Overview, has been reorganized fairly significantly, and several new descriptions were added for predictive, iterative, incremental, and adaptive (agile) life cycles.

In Chapter 3, Processes, PMI clarified and made the use of three terms more consistent. Read the definitions of work performance data, work performance information, and work performance reports. Then look at how they are being more consistently used throughout the *PMBOK® Guide*.

In Chapter 4, Integration, PMI now differentiates between the project management plan and the project documents. Project documents are not intended to be part of the project management plan. The project management plan should stay focused on defining the project management process for the project.

Chapter 5, Scope Management, added a new process called Plan Scope Management. A key highlight in this chapter is that an emphasis was changed in the Validate Scope process; the expectation is not just that deliverables will be accepted, but that they must add value and fulfill project objectives.

Chapter 6, Time Management, added a new process with a name similar to the process added in the chapter on Scope Management, Plan Schedule Management. Additional clarification is provided on the differences between management reserves and contingency reserves in planning schedules, there is a discussion of the difference between resource leveling and resource smoothing, and agile concepts are introduced.

Chapter 7, Cost Management, also has a new process, Plan Cost Management. In addition, this knowledge area has been updated to reflects the changes made in the *Practice Standard for Estimating and Practice Standard for Earned Value Management*, Second Edition, and clarifies the differences between management reserves and contingency reserves as it relates to cost.

Chapter 8, Quality Management, has had limited changes; however, several additional widely used quality models and quality tools are referenced showing the relationship between the 5 process groups and these models.

In Chapter 9, Human Resources Management, PMI has expanded on the benefits and disadvantages of virtual teams.

For Chapter 10, Communications Management, much content was moved to the stakeholder management knowledge area, and only a few minor changes were made to specific process names.

Chapter 11, Risk Management, introduces the risk profile and the terms risk attitude, risk appetite, risk tolerance, and risk thresholds. In addition, the term positive risk has been changed to opportunity risk.

For Procurement Management, Chapter 12, the only significant change has been changing a process name to Control Procurements.

ACKNOWLEDGMENTS

Diane and Janice would like to recognize the hard work and contributions of the many individuals who helped us make this book a success:

Steve Buda and his team at J. Ross Publishing for their guidance and patience; Jessica Haile for her diligent work as our editor; Michael Graupner for his never-ending encouragement and feedback to ensure our materials are better than the rest; and Scott McQuigg for his technical expertise in transforming our CD-ROM to a state-of-the-art online testing tool.

Special thanks also to Kristine Munson of the PMI-Orange County Chapter, various PMI chapters, and our students across the country for using our materials and providing invaluable feedback.

There are certainly others, whose names are not here but who contributed as participants in our workshops, challenging us to defend our statements and the answers to the PMP questions we wrote. Each course offering enhanced the material that has become this book. To each and every one of you, a million thanks from the bottom of our hearts. We could not have done this without you.

Congratulations! Your curiosity about the project management profession and what is involved in attaining the elite status of a certified Project Management Professional (PMP) will lead you to many achievements. As you may know, the *Guide to the Project Management Body of Knowledge* (*PMBOK® Guide*), which is published by PMI, is revised every four years. The revision of the *PMBOK® Guide* is followed by corresponding changes to the PMP exam questions. The current edition of the *PMBOK® Guide* is the 2012 edition (Fifth Edition), and it is often referenced as the *PMBOK® Guide* 2012. This study guide is closely related to the *PMBOK® Guide* and has been revised to match the *PMBOK® Guide* 2012. Before you begin, make sure you know what version of the *PMBOK® Guide* you need to study and determine your target time for taking the PMP exam. However, if your purpose is simply to find out more about the project management disciplines and the exam process, then any version will provide you with valuable information.

Our purpose in creating this study guide, *Achieve PMP® Exam Success: A Concise Study Guide for the Busy Project Manager*, is to provide you with a consolidated source of material that, used together with the material contained in the *PMBOK® Guide* and your experiences as a project manager, should be all you need to pass the PMP exam. To help you succeed and to make effective use of your study time, the chapter topics match the chapters in the *PMBOK® Guide* and include material on PMI's professional responsibility performance domains. Each chapter contains a series of sample exam questions and, where appropriate, a hands-on exercise or two.

We have structured each of the chapters to present a list of the things you need to know to pass the exam. This list is based on 1) our personal experiences in preparing for and taking the exam, 2) our experiences in helping others prepare for the exam, and 3) what our students and workshop participants have told us has been useful in helping them prepare for and pass the exam.

We have NOT included in this study guide all the inputs, outputs, tools, and techniques described in the *PMBOK® Guide*. PMI has already done that, so you MUST go through the *PMBOK® Guide* in detail to become familiar with the deliverables, tools, and techniques of the many processes.

Our objective in providing you with this study guide is to add value to what is described in the *PMBOK® Guide* by offering explanations and examples, as well as going beyond the *PMBOK® Guide* to topics that reinforce these concepts and improve project manager competency.

The *PMBOK® Guide*, by design, is not meant to provide detailed step-by-step instructions for the project manager in a project setting. We hope you will find that this study guide provides you the information needed to understand the concepts of the *PMBOK® Guide* and enable you to use its tools, techniques, and methods in real-world environments.

Remember also that the exam tests your knowledge of generally accepted project management processes. Your particular industry and/or area of specialization will have different ways of doing the same thing, but PMI is administering the test, so you need to know "the PMI way."

We have included an assessment test in Chapter 1 that you should take before you continue on to further chapters. It will help you focus on those areas of your knowledge and experience that are the weakest and save you time by allowing you to skim over the areas in which you are already knowledgeable. These assessment questions may also help you decide if you need to take a course on project management principles or a PMP exam preparation course before attempting the exam. Chapter 14 is a final exam to help you confirm your understanding of the material.

There are several other ancillary sources that PMI produces and that we've incorporated into this book—the *Code of Ethics and Professional Conduct*, the *PMP Exam Content Outline*, and *Appendix X-3* from the *PMBOK® Guide*. PMI is highlighting, in this edition, the importance of following the code of ethics and enhancing a project manager's interpersonal skills.

You will find at the end of many of the chapters of this study guide some guidance to improving a project manager's interpersonal skills. These skills are appropriate to all knowledge areas but have some particular importance to the knowledge area being discussed. Take them to heart; there will be exam questions that incorporate your understanding of these skills.

PMI also administers a Certified Associate in Project Management (CAPM®) exam. This is a shorter exam that has a lower experience requirement for an application to be approved. Use our *Achieve CAPM® Exam Success: A Concise Study Guide and Desk Reference* to improve your project management skills and to help you pass the CAPM exam.

For all purchasers of our book, you can download instructions on accessing an online assessment tool from the **WAV files.** Armed with the instructions and your unique serial number printed on the front cover of this book, you will have unlimited access to both 200 question practice exams and 25 question knowledge area exams for a 45-day period. A key to success in passing the PMP exam is taking several practice exams, and in particular, automated exams such as the ones we provide.

Finally, we would like to emphasize that you should by no means assume that study-ing this book replaces reading the *PMBOK® Guide*. Instead, both books should be used together. We suggest the following method of study:

- Start by doing an overview of a chapter in the *PMBOK® Guide* (paging through the chapter to get a big picture of what the chapter is about and how it is organized). This should take about two or three minutes.
- Do the same with the equivalent chapter in this study guide. Note that Chapters 1 and 2 of the *PMBOK® Guide* are combined into Chapter 2 of this study guide.
- Now, read the *PMBOK® Guide* chapter carefully, asking yourself "What do I need to learn from what I am reading?"
- Next, read the same study guide chapter carefully to find the tips and important points to learn.
- Then reread the *PMBOK® Guide* chapter in chunks, referring back to the study guide. Note concepts in the study guide that go beyond the *PMBOK® Guide*.
- Make notes or flash cards to help you remember essential information. Use these notes later to test yourself so you can narrow your focus on the information you may need to revisit.
- Next, do the sample exam questions in the study guide. If there are any concepts that you are weak in, you may want to read up on them by accessing the related reference material and practicing with additional exam questions.
- Use the online test bank to perform practice tests by simulating actual 200 question exams or arranging questions by knowledge area.
- Finally, reread the *PMBOK® Guide* chapter, this time very quickly, so you end with the overall picture rather than being buried in details.
- Then you're ready to continue this process with the other chapters.

Good studying and good luck with building your project management skills and knowledge. We look forward to hearing from you and celebrating your achievement. News of your success as well as any suggestions or comments on our study guide can be sent to us by email to info@cpconcepts.net.

Diane Altwies and Janice Preston

AUTHORS' INFORMATION

Core Performance Concepts, Inc.

Core Performance Concepts Inc, a training, curriculum, and services provider, was established in 2005 by Diane C. Buckley-Altwies and Janice Y. Preston. Answering the need for high-quality courses, curriculum, and organizational transformation services for adult learners, Core Performance Concepts provides turnkey solutions for:

- Leadership
- Project Management (including Agile methodologies)
- Process Improvement (i.e. Six Sigma, Lean)
- Business Analysis

Core Performance Concepts believes that organizations can ONLY succeed through the people they hire. Having a knowledgeable and effective workforce that understands the organizational strategy and demonstrates critical skills will enable success for any business.

Our hands-on, interactive curriculum—whether on ground or online—provides the student the necessary knowledge to prove competency in the workplace.

With over 50 years combined professional and corporate experience delivering strategy for organizations, we understand the importance of good fundamental knowledge and efficient processes and strive to help organizations succeed.

Diane Altwies, MBA, PMP, has been managing software development projects for nearly 30 years as a program manager or project manager in the insurance, financial services, and healthcare industries. She is the CEO of Core Performance Concepts, Inc. She continues to teach and consult for organizations on various program management, project management and business analysis topics and develops advanced courseware topics for project managers. She is a frequent speaker at professional meetings and symposia across the country. She is a Fellow of PMI Orange County. She has an MBA in Finance and Marketing and a BA in Production Management, both from the University of South Florida. In addition to this book, she has co-authored two others: *Program Management Professional: A Certification Study Guide with Best Practices for Maximizing Business Results* and *Achieve CAPM® Exam Success: A Concise Study Guide and Desk Reference.*

Janice Preston, MBA, CPA, PMP, has been managing projects for more than 25 years in industries as diverse as real estate, finance, healthcare, and technology. For more than 15 years, she has developed course curricula in project management and has been responsible for creating several project management certificate programs at leading universities. She writes and speaks on many project management topics, including team leadership, communication skills, earned value, cost control, and procurement. She is considered an expert in the field of risk management and has consulted on updates to the *PMBOK® Guide*'s risk management knowledge area. Janice is a Fellow of PMI Orange County, has an MBA in Finance and Accounting from the University of Missouri, a BA in Education from the University of Central Florida, and is the co-author of *Achieve CAPM® Exam Success: A Concise Study Guide and Desk Reference.*

This book has free material available for download from the
Web Added Value™ resource center at *www.jrosspub.com*

At J. Ross Publishing we are committed to providing today's professional with practical, hands-on tools that enhance the learning experience and give readers an opportunity to apply what they have learned. That is why we offer free ancillary materials for download on this book and all participating Web Added Value™ publications. These online resources may include interactive versions of material that appears in the book or supplemental templates, worksheets, models, plans, case studies, proposals, spreadsheets, and assessment tools, among other things. Whenever you see the WAV™ symbol in any of our publications, it means bonus materials accompany the book and are available from the Web Added Value Download Resource Center at www.jrosspub.com.

Downloads available for *Achieve PMP® Exam Success: A Concise Study Guide for the Busy Project Manager, 5th Edition*, include:

• Directions for how to use the online test bank with your unique serial number found on the inside front cover of this book
• A flashcard study aid of key terms and concepts
• A self-study exercise on understanding the interdependencies of all 47 processes defined in the *PMBOK® Guide*
• A training aide for better understanding process interdependencies

CHAPTER 1 | **STUDY TIPS &**
ASSESSMENT EXAM

WHAT IS A PROJECT MANAGEMENT PROFESSIONAL (PMP)?

A PMP is a project management practitioner who:
- Has demonstrated a professional level of project management knowledge and experience by supporting projects using project management tools, techniques, and methodologies
- Has at least 4,500 hours of experience as a project manager
- Has completed 35 hours of formal project management training
- Has passed a computer-based exam administered by the Project Management Institute (PMI)

PMP EXAM SPECIFICS

The exam has the following characteristics:
- It assesses the knowledge and application of globally-accepted project management concepts, techniques, and procedures
- It covers the five **performance domains** detailed in the *PMI PMP Examination Content Outline— June 2015*
- It covers the ten **knowledge areas** and the five **process groups** as detailed in the Project Management Body of Knowledge Guide (*PMBOK® Guide*)
- It contains 200 multiple-choice questions
- It includes 25 questions that are "pre-exam" questions being field-tested by PMI that do not affect your exam score
- It takes up to four hours

Table 1-1 on the following page summarizes the distribution of exam questions in each of the five performance domains based on the *PMI PMP Exam Content Outline—June 2015*. Note that the information in this chart may change from time to time. Consult the PMI website (www.pmi.org) for the most current information.

Table 1-1

Performance Domain	% of exam	# of questions
1. Initiating	13%	23
2. Planning	24%	42
3. Executing	31%	54
4. Monitoring & Controlling	25%	44
5. Closing	7%	12

STUDY TIPS

Below are a list of recommended study tips from project managers who have successfully passed the PMP exam. Think about each one and determine which suggestions are the best for your learning style.

- Develop a plan for studying; see page 1-17 for a sample study plan utilizing the *PMBOK® Guide* and this study guide
- Follow the plan on a daily or weekly basis; it is important for you to commit to studying
- Plan your study sessions with time limitations
- Study during a time of day when you are most alert
- Vary tasks and topics during lengthy study periods
- Find one special place for studying and use it only for that
- Eliminate distractions
    - If daydreaming, walk away
    - Take brief breaks (5 to 10 minutes) after about 50 minutes of study
- Use practice exams
    - Your goal should be consistently achieving 80% correctness
- Learn the *PMBOK® Guide* definitions
- Understand the big concepts first
    - Try to put the concepts in your own words
    - Try to apply concepts to your own experiences (remember your experiences may not reflect the *PMBOK® Guide* approach)
- Memorize
    - Important people and their contributions to project management
    - Formulas

1

- Processes and their order
- Inputs, tools and techniques, and outputs of each process
- Facilitate memorization by using tools like mnemonics
- Prepare for exam day
  - Get a good night's rest
  - Avoid last-minute cramming
  - Have a good breakfast
  - Leave books at home
  - Go with a positive attitude
  - Get to the exam site EARLY

EXAM TIPS

The exam tests your knowledge of the *PMBOK® Guide* by asking you many questions on definitions and inputs, tools and techniques, and outputs, and it has many situational questions that determine how well you apply *PMBOK® Guide* concepts to real-life situations. Most people can succeed if they follow these simple steps on test day.

- Use the 15 minute tutorial time to do a brain dump on the items you have memorized
- Relax before and during the exam
  - Take deep breaths
  - Stretch about every 40 minutes
  - If you get nervous, try to relax
  - Give yourself a goal and reward yourself
  - Resist the urge to panic
- Read each question carefully
- Be especially alert when double negatives are used
- Reread ALL questions containing negative words such as "not," "least," or "except"
- If a question is long and complex, read the final sentence, look at the answer choices and then look for the subject and verb
- Check for qualifying words such as "all," "most," "some," "none," "highest-to-lowest," and "smallest-to-largest"

- Check for key words such as "input," "output," "tool," "technique," "initiating," "planning," "executing," "monitoring and controlling," and "closure"
- Decide in your mind what the answer should be, then look for the answer in the options
- Reread the questions and eliminate answer choices that are NOT correct
- The correct answer, if it's not simply a number, will include a PMI term
- Make sure you look at ALL the answer choices
- Mark questions to come back to

TIME MANAGEMENT DURING THE EXAM

- Keep track of time (you have approximately 1 minute and 15 seconds for each question)
- Set up a time schedule for each question
- Allow time for review of the exam
- To stay relaxed, keep on schedule
- Answer all questions in order without skipping or jumping around
- If you are unsure, take a guess and mark the question to return to later; do not linger
- For questions involving problem solving:
    - Write down the formulas before solving
    - If possible, recheck your work in a different way (for example, rationalize)
- Subsequent questions may stimulate your memory and you may want to reevaluate a previous answer
- A lapse in memory is normal
- You will not know all the answers
- Take your time
- Do not be in a rush to leave the exam
- Before turning in the exam, verify that you have answered all questions

FAQS ABOUT THE EXAM

- Can you bring materials with you?
  NO
- What is the physical setting like?
  It is a small room or cubicle with a computer, chair, desk, and trash can
- Can you take food or drink into the exam area?
  NO food or drink is allowed
- Can you take breaks during the exam?
  YES; you can go to the restroom; your clock is ticking all the time, so you need to determine if you have time and need to take a break to clear your mind
- What are the time constraints?
  You have 4 hours (with an additional 15-minute tutorial and 15-minute survey)
- Are the exam questions grouped by knowledge area such as scope, time, and cost?
  NO; the 200 questions are randomly scattered across the process groups and knowledge areas
- Can you take paper and pen into the exam area?
  NO; pencils and paper are supplied
- Can you see both the question and the answer choices on the same screen?
  YES
- Is there a way to mark out or eliminate answer choices that you immediately know are not correct?
  NO; you can work only on a piece of scratch paper
- Is there a way to mark questions you are doubtful of?
  YES
- When you are done, can you review the exam?
  YES
- Can you review just the questions you marked as doubtful?
  YES

- Do you get immediate exam results?
  YES, if you are taking an online exam; after you are done and hit the SEND button; the computer will ask if you are sure, and after you hit SEND again, you will fill out an online evaluation of the exam process consisting of about ten questions; a testing center staff person will then give you a detailed report of your results

MEMORIZATION TIPS FOR PERFORMANCE DOMAINS, PROCESS GROUPS, KNOWLEDGE AREAS, AND PROCESSES

As stated in the *PMI PMP Exam Content Outline—June 2015*, PMI defines the field of project management as consisting of five **performance domains**:
- Initiating
- Planning
- Executing
- Monitoring and Controlling
- Closing

Each domain contains tasks and the knowledge and skills which are required to competently perform these tasks. There are also cross-cutting knowledge and skills, which are used in multiple domains and tasks.

In alignment with the five performance domains are the five **process groups**. Each process group contains two or more processes. The process groups with their corresponding process counts are:
- Initiating (2)
- Planning (24)
- Executing (8)
- Monitoring and Controlling (11)
- Closing (2)

This yields a total of 47 processes. The process groups are discussed in detail in Chapter 3 of this study guide.

1

There are ten **knowledge areas**. Each of the processes, in addition to belonging to a process group, also belongs to a knowledge area. The knowledge areas with their corresponding process counts are:
- Integration (6)
- Scope (6)
- Time (7)
- Cost (4)
- Quality (3)
- Human Resources (4)
- Communications (3)
- Risk (6)
- Procurement (4)
- Stakeholder Management (4)

The individual processes are discussed in the knowledge area Chapters 4 through 13.

Table 3-1 of the *PMBOK® Guide* has a comprehensive chart that cross-references the individual processes, knowledge areas, and process groups.

Many people like to use creative phrases that jog the memory to remember lists and sequences. Examples of memorable phrases follow for the five process groups and the ten knowledge areas.

Memorization Tip for the Five Process Groups:

Henry **I**nitiated a committee named **PEMC**o to **C**lose down the railway line.
1. Initiating
2. Planning
3. Executing
4. Monitoring and Controlling
5. Closing

EXAM TIP
Come up with your own creative phrases to remember the processes in each of the ten knowledge areas.

1

Memorization Tip for the Ten Knowledge Areas:

I've Seen That the Cost of Quality Has CRitical importance to my Project's Success.
1. Integration
2. Scope
3. Time
4. Cost
5. Quality
6. Human Resources
7. Communications
8. Risk
9. Procurement
10. Stakeholder Management

You may want to devise your own memorization tips for processes in each of the knowledge areas. Here is an example of a **memorization tip for the Project Scope Management processes**:

Planning Scope Means you Create Real Direction, a Sense of a Can-Win attitude and a ViSion for Continued Success.
1. Plan Scope Management
2. Collect Requirements
3. Define Scope
4. Create WBS
5. Validate Scope
6. Control Scope

FORMULAS, EQUATIONS, AND RULES

Some formulas, equations, and rules must be memorized to answer exam questions effectively. The most important items to remember are listed here. Most of these are discussed in more detail in the following chapters.

1. Project Network Schedules

Network schedules are created after duration estimates and the relationships between the work packages have been determined. Following a path(s) from left to right makes a forward pass.

- **Forward pass**
  - Yields early start (ES) and early finish (EF) dates
  - Early finish = early start + duration
  - RULE: If there are multiple predecessors, use LATEST EF to determine successor ES

After all paths have been given their forward path, they are traversed from right to left to make a backward pass.

- **Backward pass**
  - Yields late start (LS) and late finish (LF) dates
  - Late start = late finish – duration
  - RULE: If there are multiple successors, use EARLIEST LS to determine predecessor LF

Once the forward and backward passes have been completed, the total float for the node can be calculated by:

- Total float = late finish – early finish

1

2. Normal Distribution

The normal distribution, commonly known as the bell curve, is a symmetrical distribution, as shown in Figure 1-1. Each normal curve can be distinctly described using the mean and sum of the values. The possibility of achieving the project objective in the mean time or cost is 0%, with a 50% chance of falling below the mean and a 50% chance of exceeding the mean. Adding one or more standard deviations (σ) to the mean increases the chances of falling within the range. The probability of falling within 1σ, 2σ, or 3σ from the mean is:
- $1\sigma = 68.27\%$
- $2\sigma = 95.45\%$
- $3\sigma = 99.73\%$

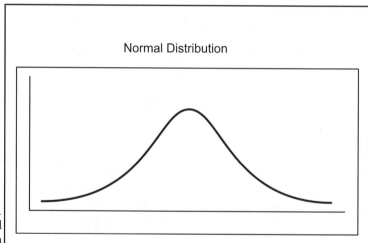

Figure 1-1
Normal
Distribution

1

3. Triangular Distribution

When there are three possible values, each of which is equally likely, the distribution takes on the shape of a triangle, as shown in Figure 1-2.

- With A = lowest value, B = highest value, and M = most likely value, variance for a task (V) (variance is not on the exam) is
  - $V = [(A - B)^2 + (M - A)(M - B)] \div 18$
- Mean (μ)
  - $\mu = (A + M + B) \div 3$
- Standard deviation (σ)
  - $\sigma = \sqrt{V}$

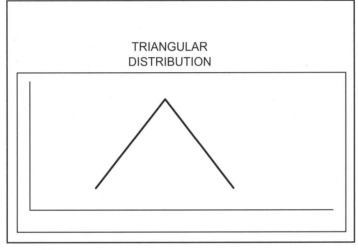

TRIANGULAR
DISTRIBUTION

Figure 1-2
Triangular
Distribution

4. Weighted-Average or Beta/PERT Distribution

The beta distribution is like the triangular distribution except more weight is given to the most likely estimate. This may result in either a symmetrical or an asymmetrical (skewed right or skewed left) graph. An asymmetrical graph is shown in Figure 1-3.

- Where O = optimistic estimate, ML = most likely estimate, and P = pessimistic estimate, variance for a task (V) is:
  - $V = \sigma^2$
- Mean (μ)
  - $(\mu) = (O + 4ML + P) \div 6$
- Standard deviation (σ)
  - $\sigma = (P - O) \div 6$

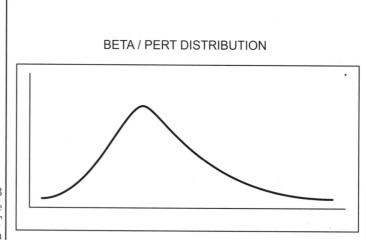

BETA / PERT DISTRIBUTION

Figure 1-3
Weighted-Average
or Beta/PERT
Distribution

5. Statistical Sums

- The project mean is the sum of the means of the individual tasks: $\mu_p = \mu_1 + \mu_2 + \ldots + \mu_n$
- The project variance is the sum of the variances of the individual tasks: $V_p = V_1 + V_2 + \ldots + V_n$
- The project standard deviation is the square root of the project variance: $\sigma_p = \sigma = \sqrt{V_p}$

1

6. Earned Value Management

Earned value management is used to monitor the progress of a project and is an analytical technique. It uses three independent variables:

- **Planned value (PV)**: the budget or the portion of the approved cost estimate planned to be spent during a given period
- **Actual cost (AC)**: the total of direct and indirect costs incurred in accomplishing work during a given period
- **Earned value (EV)**: the budget for the work accomplished in a given period

These three values are used in combination to provide measures of whether or not work is proceeding as planned. They combine to yield the following important formulas:

- **Cost variance (CV)** = $EV - AC$
- **Schedule variance (SV)** = $EV - PV$
- **Cost performance index (CPI)** = $EV \div AC$
- **Schedule performance index (SPI)** = $EV \div PV$

Positive CV indicates costs are below budget. Positive SV indicates a project is ahead of schedule.

Negative CV indicates cost overrun. Negative SV indicates a project is behind schedule.

A CPI greater than 1.0 indicates costs are below budget. An SPI greater than 1.0 indicates a project is ahead of schedule.

A CPI less than 1.0 indicates costs are over budget. An SPI less than 1.0 indicates a project is behind schedule.

1

7. Estimate at Completion

An **estimate at completion (EAC)** is the amount we expect the total project to cost on completion and as of the "data date" (time now). There are four methods listed in the *PMBOK® Guide* for computing EAC. Three of these methods use a formula to calculate EAC. Each of these starts with AC, or actual costs to date, and uses a different technique to estimate the work remaining to be completed, or ETC. The question of which to use depends on the individual situation and the credibility of the actual work performed compared to the budget up to that point.

- A **new estimate** is most applicable when the actual performance to date shows that the original estimates were fundamentally flawed or when they are no longer accurate because of changes in conditions relating to the project:
  - $EAC = AC + \text{New Estimate for Remaining Work}$

- The **original estimate** formula is most applicable when actual variances to date are seen as being the exception, and the expectations for the future are that the original estimates are more reliable than the actual work effort efficiency to date:
  - $EAC = AC + (BAC - EV)$

- The **performance estimate low** formula is most applicable when future variances are projected to approximate the same level as current variances:
  - $EAC = AC + (BAC - EV) \div CPI$
  A shortcut version of this formula is:
  - $EAC = BAC \div CPI$

- The **performance estimate high** formula is used when the project is over budget and the schedule impacts the work remaining to be completed:
  - $EAC = AC + (BAC - EV) \div (CPI)(SPI)$

1

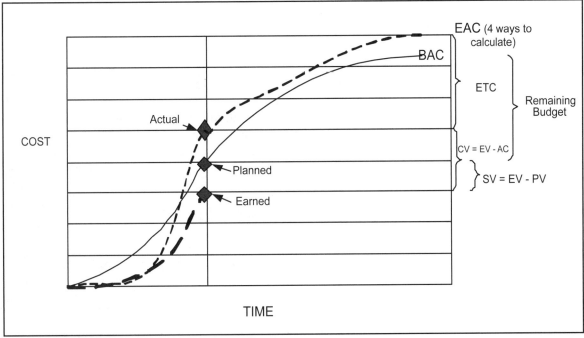

Formulas to be used with Figure 1-4 above include:
- CPI = EV ÷ AC
- SPI = EV ÷ PV

Figure 1-4
Earned Value
S-Curve

8. Remaining Budget

- RB = Remaining PV
       or
- RB = BAC – EV

9. Budget at Completion

- BAC = the total budgeted cost of all approved activities

10. Estimate to Complete

The estimate to complete (ETC) is the estimate for completing the remaining work for a scheduled activity. Like the EAC formulas above, there are three variations:
- ETC = an entirely new estimate
- ETC = (BAC − EV) when past variances are considered to be atypical
- ETC = (BAC − EV) ÷ CPI when prior variances are considered to be typical of future variances

11. Communications Channels

- Channels = [n(n − 1)] ÷ 2
  where "n" = the number of people

12. Rule of Seven

In a control chart, the "rule of seven" is a heuristic stating that if seven or more observations occur in one direction either upward or downward, or a run of seven observations occurs either above or below the mean, even though they may be within the control lines, they should be investigated to determine if they have an assignable cause. The reason for this rule is that, if the process is operating normally, the observations will follow a random pattern; it is extremely unlikely that seven observations in a row would occur in the same direction above or below the mean.

The probability of any given point going up or down or being above or below the mean is 50-50 (i.e., 50%). The probability of seven observations being consecutively in one direction or above or below the mean would be calculated as 0.50^7, which equals 0.0078 (i.e., much less than 1%).

SAMPLE STUDY SCHEDULE

Week/Day	Knowledge Area/ Domain	*PMBOK®* *Guide* Chapter	Study Guide Chapter	Goals
1	PM Overview	1 and 2	2	40 questions
2	Processes	3 and Annex A1	3	40 questions
3	Integration	4	4	40 questions
4	Scope	5	5	40 questions
5	Time	6	6	40 questions
6	Cost	7	7	40 questions
7	Quality	8	8	40 questions
8	Human Resources	9	9	40 questions
9	Communications	10	10	40 questions
10	Risk	11	11	40 questions
11	Procurement	12	12	40 questions
12	Stakeholder Management	13	13	40 questions
13	Review	1 to 13	2 to 13	100 questions
14	Review	1 to 13	2 to 13	100 questions
15	Practice	1 to 13	2 to 13	200 questions

SAMPLE ASSESSMENT EXAM

Readers should give themselves 62 minutes to answer these 50 sample exam questions. This timing is similar to the average time per question used by PMI in the actual PMP exam.

1. You are assigned as a new project manager of a project that is 40% complete. You are assigned due to the departure of the prior project manager. You review the current issues logs and meeting minutes. You determine that there has been insufficient communication with the stakeholders, and you immediately put in place:

   a) A lessons learned knowledge base
   b) A weekly face-to-face status meeting with key stakeholders until key issues have been resolved
   c) A quality audit
   d) A formal project review

2. You are a project manager on a virtual team. Each week you have a teleconference call with all the team members. During the conference call you hear someone on the call "typing" on his or her computer. This typing is an example of _____ in the communication model.

   a) A message
   b) Noise
   c) Encoding
   d) Decoding

Notes:

3. You are the project manager for a business process improvement project for a strategic business process that is 50% complete. As a major milestone, a test was performed to determine if the new process being developed may actually make the production user less productive, if implemented. One way to analyze the results of this test is to perform:

   a) A root cause analysis
   b) Quality audits
   c) A defect repair review
   d) A Pareto diagram

4. A breakdown in communication _____ impacts the project.

   a) Usually positively
   b) Sometimes
   c) Always
   d) Usually negatively

5. How do activities differ from milestones?

   a) Activities are the lowest level in the WBS; milestones are high level with short durations
   b) Activities have duration; milestones don't because they represent a point in time
   c) Activities can be used to highlight a significant point in the project; milestones are scheduling units
   d) Activities have abbreviated descriptions; milestones must be well defined

Notes:

6. You are the project manager on a construction project that is 50% complete. At this point the CPI is 1.12 and the SPI is 1.05. Total earned value to date is $6,300,000, and the original budget was $12,600,000. All of the following are true except:

   a) Actual costs are $6,300,000
   b) EAC could be $11,925,000
   c) Actual costs are $5,625,000
   d) EAC could be $11,250,000

7. The act of collecting and distributing status reports, progress measurements, and forecasts is part of:

   a) Communication technology
   b) Communication methods
   c) Information management systems
   d) Performance reporting

8. You are the project manager for a business process improvement project for a strategic business process. Several new business process needs have been identified as planning packages in the WBS, and as part of the initiation process for the project, a sub-team has been assigned the responsibility to detail out the deliverable expectations of these new planning package. This process is known as:

   a) Analogous estimating
   b) Bottom-up estimating
   c) Discovery
   d) Decomposition

9. You are the project manager on a new project in which 10 of the 12 team members have worked together on prior similar project. Senior management wants a quick estimate on the overall cost of the project. A good tool that can be used in making this estimate is:

   a) Analogous estimating
   b) Bottom-up estimating
   c) Three-point estimating
   d) Resource leveling

Notes:

10. You are the project manager on a project in which two of your key stakeholders have different concerns and expectations on the objectives of the project. Your job is to:

    a) Assess the situation to balance expectations of the key stakeholders
    b) Prioritize the objectives and manage your customers' expectations
    c) Have the stakeholders prioritize their objectives and focus on the most common ones
    d) Talk to senior management and determine which objectives best fit the overall business goals of the organization

11. As the project manager, Jacob identifies the possible departure of a key team member who has a special expertise that is needed for the project. If that team member is promoted, this would be considered a risk:

    a) Category
    b) Event
    c) Trigger
    d) Quantification

12. As a project manager on a construction project, you experienced an abnormal number of vendor issues on your project. In order to prevent the situation from happening on future contracts, you ask for a:

    a) Checklist analysis
    b) Reserve analysis
    c) Procurement audit
    d) Risk audit

13. A requirements traceability matrix is a table that links requirements to their origin and can be used to control:

    a) Communications
    b) Stakeholders
    c) Deliverables
    d) Schedule

Notes:
_____
_____
_____
_____
_____
_____
_____
_____
_____
_____
_____
_____
_____
_____
_____
_____
_____
_____
_____
_____
_____
_____
_____
_____

14. Your virtual team has developed bad habits, such as being late to meetings, multi-tasking during meetings, and carrying on side discussions. What should the project manager do to reduce or prevent such problems?

    a) Conduct a training session on soft skills to build empathy and influence
    b) Have a team building session to help members get to know each other better
    c) Bring in the team for a face-to-face meeting and discuss behavior expectations
    d) Have the team agree to ground rules with clear expectations for meetings

15. The subsystem of formal documented procedures to track and manage characteristics and changes in a deliverable is known as the:

    a) Contingency reserves system
    b) Procurement administration system
    c) Configuration management system
    d) Integrated change control system

16. In which contract type does the seller have to include any contingencies for risk and the seller's profit as part of the total contract price?

    a) Cost reimbursable
    b) Firm fixed price
    c) Cost plus incentive fee
    d) Cost plus penalty fee

17. How is earned value analysis useful in controlling risk?

    a) A CPI below 1.0 may indicate a threat that has occurred
    b) Schedule variance that is positive may indicate a threat that has occurred
    c) Cost variance that is negative means you are under budget and things are okay
    d) An SPI greater than 1.0 means you are behind schedule and should investigate

Notes:

18. You are a project manager newly assigned to a critical project for the organization. After reviewing the project charter and evaluating the needs of the project, it is clear to you that Nikhil would be a critical resource on the project because of his product knowledge and customer relationship. As part of your resource planning process, you _____ Nikhil's boss to allocate his time to the project.

   a) Direct
   b) Negotiate with
   c) Compromise with
   d) Demand of

19. A dependency that is inherent in the nature of the work being performed is a (an) _____ dependency.

   a) Discretionary
   b) External
   c) Contractual
   d) Mandatory

20. As the project manager on a software development project, you schedule phase-end lessons learned reviews with the entire project team. The results of the session will be:

   a) Distributed to all internal stakeholders regardless of the results
   b) Kept by the project manager for future projects
   c) Shared only with the team members as a learning experience
   d) Used to identify who on the team was not performing up to standards

Notes:

21. You are working on a project that was recently initiated to enhance the community parks within the state. Your government agency is highly influenced by legislative changes. A new law is scheduled to be implemented within the first three months of the project that will require changes to security in all parks. You should:

    a) Do nothing because these changes weren't part of the initial scope
    b) Submit a change request to the change control board to be voted on
    c) Submit a change request to the change control board as a required change
    d) Include the changes in your current project scope

22. The difference between grade and quality is that:

    a) Grade is looking at the attributes of the product, whereas quality is concerned with how well the product meets requirements
    b) Grade is fitness-for-use, while quality is concerned with how well the product meets requirements
    c) Grade is the cost to confirm to specific requirements; quality is the cost of non conformance
    d) Grade and quality are actually synonymous and have no real difference on projects

Notes:

23. Given the information provided below regarding an upcoming party, which option should be chosen if the goal is to maximize revenue?

   Notes:

   • If the party is held inside, 100 people can attend at $10 each
   • If the party is held outside, 200 people can attend at $10 each
   • There is a 40% chance of rain on the date of the party
   • If it rains and the party is held inside, only 80 people will attend
   • If it rains and the party is held outside, only 100 people will attend

   a) Held inside because expected monetary value is $1,800
   b) Held inside because expected monetary value is $920
   c) Held outside because the expected monetary value is $3,000
   d) Held outside because the expected monetary value is $1,600

24. Determining how to deal with each stakeholder is a key benefit of which process?

   a) Identify Stakeholders
   b) Plan Stakeholder Management
   c) Manage Stakeholder Engagement
   d) Control Stakeholder Engagement

21, B
22, A
23, C
24 C

MAX
max technical training

513.322.8888
www.maxtrain.com

25. A subcontracting vendor completes work to the requirements specified within the statement of work (SOW), but the project manager is not pleased with how the deliverable integrates with the corporate databases. In this case, the contract is considered to be:

   a) Incomplete because formal acceptance has not been provided by the buyer
   b) Complete because the contractor met the terms and conditions of the contract
   c) Complete because the contractor is satisfied and work results followed the SOW
   d) Incomplete because the specs are incorrect

26. You are a project manager on a hotel construction project. The CEO and sponsor of the project retired from the business and a new CEO has been named. One of the first things you should do as the project manager is:

   a) Introduce yourself to the new CEO and tell him or her how hardworking your team is
   b) Schedule a meeting with the CEO to review the detailed project schedule and earned value analysis that you finished last Friday
   c) Determine the communication needs of the CEO and modify the communications management plan
   d) Keep your head down and continue to perform your job

27. What is the benefit of following the communication plan to collect and distribute information?

   a) It allows an efficient flow of communications
   b) It identifies and documents the approach to communicate
   c) It enables the project manager to use interpersonal skills effectively
   d) It ensures that information is flowing to all project stakeholders

Notes:

28. Those risks that can be identified and analyzed are called:

    a) Known risks
    b) Unknown risks
    c) Risk register
    d) Risk responses

29. You are the project manager of a technology project and several questions have arisen as the technology team began the design phase. As a result, you schedule a meeting with the key stakeholders to review the issues and bring clarity to the design. This is an example of which process?

    a) Manage Stakeholder Engagement
    b) Report Performance
    c) Plan Stakeholder Management
    d) Control Stakeholder Engagement

30. Your project is nearing completion, and you have scheduled a deliverable review meeting with your customer for next week as part of the Validate Scope process. The Validate Scope process includes:

    a) Obtaining the stakeholders' formal acceptance of the project's deliverables
    b) Organizing and defining the total scope of the project
    c) Verifying the correctness of the project deliverables
    d) Assuring that all requested changes and recommended corrective actions are completed

Notes:

31. A member of your team brings ideas for enhancements to the scope of work to a team meeting. These suggestions will add work to the project that is beyond the requirements of the project charter. As project manager, you point out that only the work required for the project should be completed by the team or the project could miss its goals. You are:

a) Acting as the change control board
b) Collecting requirements
c) Verifying scope
d) Managing risk

32. As you begin implementation of your website redesign project, what enterprise environmental elements should you consider?

a) Procedures used to detect and track bugs
b) Existing capacity of the website infrastructure
c) Allowed communication methods
d) Project files from previous projects

33. One difference between project manager responsibility and PMO responsibility is that project managers:

a) Focus on specific project objectives, while PMOs manage program changes
b) Optimize the use of shared resources, while PMOs manage assigned resources
c) Manage the overall risks and opportunities, while PMOs control a project's scope, cost, and time
d) Manage the methodology and metrics used, while PMOs manage individual reporting requirements

34. A narrative description of products to be supplied by the project is the:

a) Project scope statement
b) Charter
c) Project statement of work
d) Contract

Notes:

1

35. During the closing phase of a project, employees are concerned about their next assignment in which type of organizational structure?

a) Projectized
b) Strong matrix
c) Balanced matrix
d) Functional

36. When is it critical to have procedures for identifying and documenting actions taken during the project?

a) When the project deliverables satisfy the completion criteria
b) When a project is terminated before completion
c) When you have transferred the product service into ongoing operations
d) When archiving lessons learned and other project information

37. Which technique would be most effective in increasing the effectiveness of stakeholder engagement activities as the project progresses when there are major disagreements on the deliverables?

a) Reviewing the issues log
b) Holding face-to-face meetings
c) Using the information management system
d) Using the expertise of an external consultant

38. As the project manager on a medical device project, you are asked to put together an estimate for the project. At a minimum, you would need:

a) A scope baseline and human resource management plan
b) Activity cost estimates and the project budget
c) Analogous and bottom up estimates for labor costs
d) Project management software and a vendor bid analysis

Notes:

39. The PMI *Code of Ethics and Professional Conduct* requires that project management practitioners do all of the following EXCEPT:

    a) Uphold policies and laws
    b) Provide accurate information in a timely manner
    c) Negotiate in good faith
    d) Disclose real or potential conflicts of interest

40. You are a team member on a resource constrained project, and you find out that you can get a resource to help you for three weeks, so you write up a request to the project manager because you know that contingency reserves:

    a) Were included as part of the planning process
    b) Were only for unknown unknown cost uncertainties
    c) Were not included in the project budget
    d) Were only for scheduled activities that utilize time and materials

41. As the project manager of a large residential construction project, you find that the developer has presented three new landscaping designs to the investors, even though the contract has been signed and construction has begun. In your estimate, all three designs would add $10,000 to $30,000 to the landscaping budget. You ask for a meeting with key stakeholders immediately to address this potential variance. You are exercising:

    a) Scope control
    b) Scope baselining
    c) Schedule management
    d) Cost management

Notes:

42. You have been working on your project in a foreign country. While you were there, the local team developed a prototype of your product that you are taking home. As you exited through customs, the prototype was confiscated, even though it was declared and you paid a customs fee. To retrieve the prototype you were forced to pay an additional, unwritten "customs fee." On your expense report, you:

a) Show nothing for the additional customs fee
b) Bury the customs fee in other parts of your expense report
c) Report to the sponsor on your expense report that the customs fee needs to be reimbursed
d) Include the cost of the additional customs fee since it was a prototype

43. One of your team members, Sue, is so enthusiastic that she tends to go above and beyond the requirements for any particular task assigned. In her desire to do a great job, her work tends to be delayed and takes much longer than budgeted. This has negatively impacted the overall schedule of the project. For the next work package to be assigned to Sue, the project manager should:

a) Tell Sue's manager that Sue's work is substandard
b) Talk to Sue about the delays and clearly spell out the deliverables and timelines
c) Inform your sponsor that you are removing Sue from the team
d) Discuss the issue in the weekly project status meeting and gain consensus of the entire team

Notes:

44. Michael has taken over a twelve-month project which has two months of work completed. The prior project manager left for a promotion and told Michael the team was functioning well together. During the first several team meetings, Michael noticed a significant amount of disagreement and a lack of collaboration among team members. What happened to the high-functioning team?

    a) The team was ready to be released and to move on from the work completed
    b) The team was in a phase in which team members were beginning to adjust their work habits to support the team
    c) The team enjoyed the conflict and actively tried to express the ideas and opinions of all the team members
    d) The change in project manager shifted the dynamics and caused the group to go back through the storming phase

45. Project managers must balance competing constraints such as cost, scope, time, resources, risk, and quality. How is a constraint affected if another constraint changes?

    a) There is no affect to the remaining constraints
    b) Cost is unlikely to change if scope is kept under control
    c) One or more of the other constraints is likely to change
    d) The schedule will be kept constant if it is the most important constraint

Notes:

46. As the project manager, you review a scatter diagram in which a large majority of the points are collecting close to the diagonal line. This would communicate that the:

    a) Trend is moving upward, which identifies a potential problem on the project
    b) Risk is increasing on your ability to meet the project schedule and cost
    c) Variables measured are closely related
    d) Variables are out of control

47. You are assigned as a new project manager of a project that is 40% complete. You are assigned due to the departure of the prior project manager. You review the current issues logs and meeting minutes, and you talk to the project sponsor and other key stakeholders. You are told that the business environment that you are working in is in a critical state and that it is imperative that this project be completed on time to help the organization weather a volatile market shift. You:

    a) Update the work breakdown structure for additional work
    b) Document the lessons learned knowledge base
    c) Add schedule activities to the project schedule
    d) Review the risk register for risk events that will cause a delay

48. You are the project manager in a construction project. The framing subcontractor stated earlier this week that he will be able to meet the delivery schedule. You find out from the laborers on site that the framing subcontractor has laid off half of its staff, several of whom were working on this project. When reporting the status of the project to senior management, you should report:

    a) No slip in schedule
    b) That you have contacted the subcontractor to obtain a revised status
    c) Report that the schedule is very likely to slip
    d) That a buffer task was added to the framing task

Notes:

49. As the project manager for a business process improvement project for a strategic business process, you have been given a budget of $3.0 million to deliver the project within twelve months. You were not included in the discussions determining this budget, but your boss was. The project has been identified as a key strategic initiative this year, and if the project is delivered, the organization will experience a 30% reduction in process costs. Your first action, upon receiving the budget, should be:

a) Compare the assumptions used to develop the budget to the overall scope of the project
b) Update the cost management plan for all the changes that you expect on the project
c) Review the budget in detail and and ask your boss for a revised budget for problem areas
d) Verify that there is enough budget to cover your salary

50. You are the project manager on a large government contract. You have determined that you will need to procure a major component of the project due to a lack of expertise within your existing project team. As part of the procurements process, you receive over fifty seller proposals. In order to select a seller, you decide to:

a) Review the proposals and pick the two lowest bidders
b) Call a bidders conference and ask detailed questions about each proposal
c) Review each proposal in detail and select the winner
d) Use established criteria to rank the proposals and select a seller

Notes:

SAMPLE ASSESSMENT EXAM ANSWERS

with explanations and references are in Chapter 15, Appendix A.

CHAPTER 2 | **PROJECT MANAGEMENT OVERVIEW**

2

2

2

PROJECT MANAGEMENT OVERVIEW

Chapters 1 and 2 of the *PMBOK® Guide* provide a basic structure for the field of project management. These chapters introduce project management and the context or environment in which projects operate. Together, these first two chapters contain many important definitions and concepts that must be understood before attempting the remaining chapters of the *PMBOK® Guide*.

Project management overview questions on the PMP exam mainly cover definitions, concepts, and approaches. You must be very familiar with PMI terminology. Projects, programs, project management, stakeholders, project and product life cycles, organizational structures, and influences are among the topics covered.

Things to Know

1. The various **project constraints**
2. The definition of **business value**, **enterprise environmental factors**, and **organizational process assets**
3. The difference between **project management**, **program management**, and **portfolio management** as each relates to the various knowledge areas
4. The difference between **projects**, **programs**, and **portfolios**
5. The purpose of the **project management office**
6. Project management's role in **operations management** and in **organizational strategy**
7. The **roles** and **interpersonal skills** of the project manager
8. The many **organizational cultures** and **styles**
9. The importance of **organizational communication**
10. Three primary **forms of organizational structure**
11. The value of **governance**
12. Variations in **project team** composition
13. The differences between **projects**, **products**, and their respective **life cycles**
14. The preferred use of **incremental**, **predictive**, and **adaptive life cycles**
15. The concept of the **influence curve**

> **EXAM TIP**
> Reference the Glossary of the *PMBOK® Guide* frequently to learn PMI terminology.

Key Definitions

Business value: the entire value of the business; the total sum of all tangible and intangible elements.

Colocation: project team members are physically located close to one another in order to improve communication, working relations, and productivity.

Constraints: a restriction or limitation that may force a certain course of action or inaction.

Good practice: a specific activity or application of a skill, tool, or technique that has been proven to contribute positively to the execution of a process.

Enterprise environmental factors: external or internal factors that can influence a project's success. These factors include controllable factors such as the tools used in managing projects within the organization or uncontrollable factors that have to be considered by the project manager such as market conditions or corporate culture.

Operation: ongoing work performed by people, constrained by resources, planned, executed, monitored, and controlled. Unlike a project, operations are repetitive; e.g., the work performed to carry out the day-to-day business of an organization is operational work.

Organizational process assets: any formal or informal processes, plans, policies, procedures, guidelines, and on-going or historical project information such as lessons learned, measurement data, project files, and estimates versus actuals.

Portfolio: a collection of programs, projects, and additional work managed together to facilitate the attainment of strategic business goals.

Product life cycle: the collection of stages that make up the life of a product. These stages are typically introduction, growth, maturity, and retirement.

Program: a group of related projects managed in a coordinated way; e.g., the design and creation of the prototype for a new airplane is a project, while manufacturing 99 more airplanes of the same model is a program.

Progressive elaboration: the iterative process of continuously improving the detailed plan as more information becomes available and estimates for remaining work can be forecast more accurately.

Project: work performed by people, constrained by resources, planned, executed, monitored, and controlled. It has definite beginning and end points and creates a unique outcome that may be a product, service, or result.

Project life cycle: the name given to the collection of various phases that make up a project. These phases make the project easier to control and integrate. The result of each phase is one or more deliverables that are utilized in the next few phases. The work of each phase is accomplished through the iterative application of the initiating, planning, executing, monitoring and controlling, and closing process groups.

> **EXAM TIP**
> Read the PMI *Lexicon of Project Management*. It provides the foundational professional vocabulary.

Project management: the ability to meet project requirements by using various knowledge, skills, tools, and techniques to accomplish project work. Project work is completed through the iterative application of initiating, planning, executing, monitoring and controlling, and closing process groups. Project management is challenged by competing and changing demands for scope (customer needs, expectations, and requirements), resources (people, time, and cost), risks (known and unknown), and quality (of the project and product).

2

Project management information system: the collection of tools, methodologies, techniques, standards, and resources used to manage a project. These may be formal systems and strategies determined by the organization or informal methods utilized by project managers.

Stakeholders: individuals and organizations who are involved in or may be affected by project activities. Examples of stakeholders include the project manager, team members, the performing organization, the project sponsor, and the customer. The *PMBOK® Guide* advocates that any discrepancies between stakeholder requirements should be resolved in favor of the customer. Therefore, the customer is one of the most important stakeholders in any project.

Standard: a document that describes rules, guidelines, methods, processes, and practices that can be used repeatedly to enhance the chances of success.

Subproject: a component of a project. Subprojects can be contracted out to an external enterprise or to another functional unit.

PROJECT CONSTRAINTS

Projects are often performed under many constraints that could impinge on the project's successful completion. In addition, these constraints interact and require tradeoffs or decisions that must be made to fulfill project objectives.

For example, additional scope requirements will usually mean either more time to complete those requirements or more resources to work on these requirements, thereby increasing project cost as well as creating additional project teams. These project constraints were previously known as the triple constraint.

2

The *PMBOK® Guide* has expanded the number of constraints that need to be balanced in managing a project. These constraints now include:

- Scope
- Quality
- Budget (Cost)
- Resources (Cost, Time)
- Schedule (Time)
- Risk

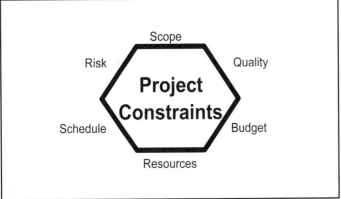

Figure 2-1
Project Constraints

If one factor changes, one or more other factors are impacted, as depicted in Figure 2-1. In addition, **enterprise environmental factors** may also constrain or limit the project team's ability to function.

BUSINESS VALUE

Every organization has value. If there were no value in a business, it would not exist. Some businesses provide commercial value, while others benefit the community or stockholders. Each organization's value is unique, just as each project is unique. For any project to be successful, the project manager and project team must understand how the project relates to the value of the business. Understanding the business value of an organization will make project managers' responses to varying situations more effective.

2

ENTERPRISE ENVIRONMENTAL FACTORS

The workplace has changed tremendously in the last two decades, forcing organizations to compete in a global economy. Various internal and external factors can and often do contribute to, or detract from, a project's success.

Project teams are now often geographically dispersed, and the colocation of project team members, although still a viable technique, is often no longer possible. With the advent of **virtual teams** (project teams that spend little or no time meeting face-to-face), the enterprise environmental factors and cultural norms, standard processes, common project management information systems, and the organization's established **communications channels** are more important than ever.

Some best practices for project teams with one or more virtual team members are:
- Use web tools for virtual meetings to facilitate communications among team members and key stakeholders
- Be conscious of different time zones and cultures
- Use a **virtual team room**, BLOGs, WIKIs, or other collaboration tools for project deliverables and work products
- Be familiar with the organizational process assets and project management information systems, and have both easily accessible
- Use a proven defined approach, and document the adaptations (tailoring) of the process to fit the needs and requirements of the project
- Have frequent and regular contact with all stakeholders (virtual or face-to-face)
- Hold regularly scheduled team meetings to define requirements, discuss issues, review deliverables, and make decisions
- Use multiple methods of communication, such as mail, email, phone calls, phone conferences, virtual meetings, face-to-face meetings when possible, and teleconferences

ORGANIZATIONAL PROCESS ASSETS

An organizational process asset can be any tangible property or resource of the organization that the project team has access to use, re-use, tailor, or modify to support the project effort. The following are examples of situations in which a project manager utilized an organizational process asset:

- Using a schedule template for an information technology project within the same organization as the starting point for the WBS development activities
- Using a prior quality control plan as the basis for another project's quality control plan
- Inserting current human resource guidelines for hiring and managing contractor resources for the project within the human resources plan
- Reviewing a prior project's lessons learned document to trigger thought and discussion on potential risks that could be encountered on the existing project

RELATIONSHIPS BETWEEN PROJECT MANAGEMENT, PROGRAM MANAGEMENT, AND PORTFOLIO MANAGEMENT

Many organizations use the terms project, program, and portfolio very loosely. For the exam, you must understand the specific definitions of each term and how they relate to one another. In addition, you must recognize that a project manager's function is very different from the function of a program manager or portfolio manager. Although each uses the same set of **knowledge areas**, these individuals have very different functions to perform during each knowledge area.

> **EXAM TIP**
> Read Table 1-1 of the *PMBOK® Guide* to understand the different functions performed by project managers, program managers, and portfolio managers within each knowledge area.

Projects, Programs, and Portfolios

Projects are unique, one-time endeavors with a defined beginning and end. They have specific objectives to fulfill, which are achieved through the coordination of interrelated tasks and activities.

Projects are not independent events within an organization. They are one piece of an overall strategic plan. The projects that an organization undertakes should facilitate the achievement of that strategic plan. They should be prioritized so that the most important projects are given every opportunity to succeed and should regularly be re-assessed as to their impact on the overall corporate vision.

A **program** is a collection of related projects that have a single focused objective. Managing projects within a program adds complexity and requires additional coordination between the projects within the program. However, program management can enhance the value of projects by coordinating seemingly independent activities. Programs may include elements of related work outside of the scope of the discrete projects in the program.

A **portfolio** is a group of projects that are coordinated so that the organization can implement its business strategy and organizational vision. The projects or programs in the portfolio may not be interdependent or directly related.

There is usually no lack of projects within an organization, but every organization has a limited amount of time, money, staff, expertise, assets, and other resources. Two aspects that are most important in choosing projects are critical, specialized resources and money. There are always more "good" projects that could be selected than there are resources.

Portfolio management is the pursuit of a balanced portfolio of projects. The balance comes from comparing several factors, which may include:

- External market-driven costs versus internal cost reduction
- Enterprise versus business unit benefit
- Research and development versus existing product lines
- Short-term versus long-term goals
- High risks versus low risks

2

Portfolio management will aid in managing **scarce resources**. It benefits business units as they plan and execute projects. It provides senior management a way to compare projects across the organization and to consider new prospects that arise during the course of business. It also assists in managing project and organization risks.

Organizations that manage portfolios of projects and programs have a greater capability to plan and predict their financial results. When projects and programs are defined in terms of their contribution to the organization, senior management makes better decisions about the mix of projects and programs and their associated values.

Portfolio management helps all levels and business units communicate, which increases the probability that the organization will have long-term financial success.

Project Management Office (PMO)

The project management office is an additional layer of organization dedicated to helping project managers. Although most often found in matrixed or projectized organizations, a project management office may exist in any type of organizational structure. Figure 2-2 shows the pros and cons of adding a PMO layer of organization.

Figure 2-2
PMO Pros and Cons

PROS	CONS
• Emphasis on project management career paths • Less anxiety among project managers about the next assignment at project completion • Centering of project management competencies • Standardization of the project management system • Centralized management	• Additional layer of hierarchy • Some of the adverse aspects of a matrix organization • All of the adverse aspects of a projectized organization • Lack of application knowledge by the project managers

2

PROJECT MANAGEMENT IN OPERATIONS MANAGEMENT AND ORGANIZATIONAL STRATEGY

The *PMBOK® Guide* emphasizes the role projects play within an organization's operations and in strategic planning, and it has also made a clear distinction between operations and project management. You must understand how projects are critical to the operations of a business in achieving organizational goals and how project management supports operations.

No organization will grow without an excellent execution of strategy. An organization chooses projects that directly deliver components of an organizational strategy.

ROLES AND INTERPERSONAL SKILLS OF THE PROJECT MANAGER

The *PMBOK® Guide* emphasizes the **responsibilities** and **competencies** necessary for project managers to succeed. Project managers must have enough knowledge to perform their function. They must be able to execute all necessary work and must do so in a professional and ethical way.

In addition, the *PMBOK® Guide* stresses the importance of interpersonal skills. It specifically lists 11 such skills that can be leveraged in the various situations a project manager will encounter. These skills are:
- Leadership
- Team building
- Motivation
- Communication
- Influencing
- Decision making
- Political and cultural awareness
- Negotiation
- Trust building
- Conflict management
- Coaching

EXAM TIP

Read the PMI *Code of Ethics and Professional Conduct*.

In this book, as we discuss each of the ten knowledge areas, we may highlight these skills and demonstrate how the interpersonal skill can be used to a project manager's advantage.

ORGANIZATIONAL CULTURES AND STYLES

Organizational cultures and styles play a large part in any organization. Every organization is different, and what works in one organization may not work in another. The project manager must be able to assess the organizational cultures and styles of both the performing organization as well as external organizations that may be interacting with the project, such as the customer or vendor.

Knowing the **vision**, **values**, **regulations**, **risk tolerance**, and **work ethic** of an organization, to name a few, will change how a project manager manages and responds to situations on the project.

ORGANIZATIONAL COMMUNICATION

Project managers are told from the very beginning that communication is a large part of their job. As the profession of project management matures and as the use of **virtual teams** grows, the complexities of communication will increase and the risks associated with **poor communication** will also increase.

It is paramount that the project manager use a variety of communication tools to ensure good and clear communications to all the project stakeholders.

FORMS OF ORGANIZATIONAL STRUCTURE

The *PMBOK® Guide* stresses the importance of organizational structures because the organizational structure will often constrain the availability of resources for a project. Become very familiar with Table 2-1 in the *PMBOK® Guide*, Organizational Influences on Projects.

Functional Organization

In a functional organization, each employee is in a hierarchical structure with one clear superior. Staff is grouped by specialty, such as accounting, marketing, or engineering. The pros and cons of a functional organization are shown in Figure 2-3 below. Included in a functional organization is the use of a **project expeditor** or a **project coordinator**.

Project expeditor (PE): the PE is a facilitator who acts as the staff assistant to the executive who has ultimate responsibility for the project. This person has little formal authority. The PE's primary responsibility is to communicate information between the executive and the workers. This type of structure is useful in functional organizations in which project costs are relatively low.

Project coordinator (PC): the PC reports to a higher level in the hierarchy and is usually a staff position. A PC has more formal authority and responsibility than a PE. A PC can assign work to functional workers. This type of structure is useful in functional organizations in which project costs are relatively low compared to those in the rest of the organization.

Figure 2-3
Functional
Organization
Pros and Cons

PROS	CONS
• Flexibility in staff use • Availability of experts for multiple projects • Grouping of specialists • Technological continuity • Normal advancement path	• Client is not the focus of activity • Function rather than problem oriented • No one fully responsible for the project • Slow response to the client • Tendency to suboptimize • Fragmented approach to the project

Matrix Organization

Understand the matrix organizations—weak, balanced, and strong—and how they differ. The pros and cons of a matrix organization are listed in Figure 2-4 below. Matrix organizations have:

- High potential for conflict
- Team members who are borrowed from their functional groups and who are therefore caught between their functional manager and their project manager (but as projects draw to a close, these team members know they have a "home" with their functional groups)
- Team members who only see pieces of the project and may not see the project to completion
- An advantage in relatively complex projects in which cross-organizational knowledge and expertise are needed
- Project managers whose authority and time on a project increases from weak matrix (lowest) to balanced matrix to strong matrix (highest)

EXAM TIP

The *PMBOK® Guide* places enormous emphasis on the social, economic, and environmental influences on projects. A key influence on the role and authority of the project manager is the organizational structure. Anyone wishing to pass the exam must understand organizational influences.

Figure 2-4
Matrix Organization Pros and Cons

PROS	CONS
• Project is the point of emphasis • Access to a reservoir of technical talent • Less anxiety about the team's future at project completion • Quick client response • Better firm-wide balance of resources • Minimizes overall staff fluctuations	• Two-boss syndrome • More time and effort needed to acquire team members • Functional managers may be reluctant to share top performers • Conflicts of authority between project manager and functional manager • Careful project monitoring required • Political infighting among project managers

Projectized Organization

In a projectized organization, team members are often colocated and the project manager has a great deal of independence and authority. Team members worry about their jobs as a project draws to a close. Figure 2-5 below shows the pros and cons of a projectized organization.

PROS	CONS
• One boss • Project manager has a great deal of independence and authority • Team members are often colocated • Team members are treated as insiders • Most resources are involved in project work	• If not tracked closely, hourly costs may become inflated while specialists are waiting between assignments or are on call • Bureaucracy, standards, procedures, and documentation may result in an abundance of red tape

Figure 2-5
Projectized
Organization
Pros and Cons

GOVERNANCE

Project governance is an **oversight function** that provides an organization's project teams direction and structure for successfully delivering projects. **Project management offices**, **program management offices**, and **portfolio management offices** are all forms of governance; however, within an organization, how each office provides value to the project teams will vary from company to company.

For the exam you should know the various elements that could be included in a governance framework.

PROJECT TEAMS

Project teams and their composition can also impact the success of a project. One of the key challenges project managers face is project staff who are part-time and not dedicated to the project. Organizational culture, project scope, and resource location are all factors in team composition.

2

Additionally, projects interact with external entities which can also impact the team composition. For the exam, you should understand how having a partnership, joint venture, consortium, or alliance could impact the overall team effectiveness and thereby change how a project manager approaches the management of the project.

PROJECT LIFE CYCLE

A project life cycle defines:
- The phases that a project goes through from initiation to closure (the *PMBOK® Guide* states that a project contains an initial phase, one or more intermediate phases, and a final phase)
- The technical work to be done in each phase
- The skills involved in each phase
- The deliverables and acceptance criteria for each phase
- How each phase will be monitored, controlled, and approved before moving to the next phase

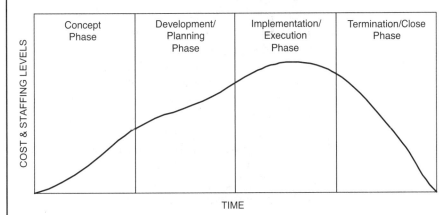

Figure 2-6
Project Life Cycle

A typical project life cycle contains the following four phases, as shown in Figure 2-6.

2

- **Starting the project** (the **concept phase**): the problem to be solved is identified. Deliverables from this phase could be:
  - Feasibility studies that clarify the problem to be solved
  - Order of magnitude forecasts of cost
  - A project charter to grant permission for the project to proceed
- **Organizing and preparing** (the **development and planning phase**): what needs to be done is identified. Deliverables created here include:
  - The scope statement
  - A work breakdown structure (WBS)
  - A schedule baseline
  - A determination of budgetary costs and a developed budget
  - The identification of resources and team members with levels of responsibility
  - A risk assessment
  - A communications management plan
  - The project management plan
  - Control systems and methods for handling change control

- **Carrying out the work** (the **implementation and execution phase**): the actual work of the project is carried out. Deliverables include:
  - Execution results for work packages
  - Status reports and performance reporting
  - Procurement of goods and services
  - Managing, controlling, and redirecting (if needed) scope, quality, schedule, and cost
  - Resolution of problems
  - Integration of the product into operations and the transferral of responsibility

- **Closing the project** (the **termination and close phase**): the product is finalized, evaluated, and rejected or accepted. Deliverables include:
  - Formal acceptance
  - Documented results and lessons learned
  - Reassignment or release of resources

2

Projects versus Products

A project is a temporary endeavor that is undertaken to create a unique product, service, or result. When the outcome of a project is related to a product, the outcome of the project could, for instance, be:
- The development of a new stand-alone product
- The addition of new functions or features to an existing product
- The development of a component or segment of a product or of an aspect of a product such as a prototype or installation at a new location

A product is an artifact that is produced and is quantifiable. It can either be an end item such as an airplane, or a software application, or it can be a component item such as an engine, or a software feature.

Relationship of Project Life Cycle to Product Life Cycle

The life cycle of a project is only one aspect of the overall **product life cycle**, as Figure 2-7 below shows. A project can be initiated to determine the feasibility of a product in the introductory stage of a product life cycle. There may be a second project to address the design and development of the product once the feasibility study has determined the viability of the product.

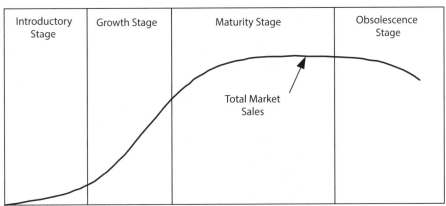

Figure 2-7
Product Life Cycle

2

The number of projects initiated to support the product life cycle will vary from organization to organization and from product to product.

Project life cycle phases and product life cycle phases are often defined similarly. For example, a project life cycle may start with a feasibility phase to determine if the project can achieve its objectives while the first phase in a product life cycle might consist of a market study to determine if the product will meet sales goals.

Phases of a product life cycle are generally performed in sequence. Although the phases in a project life cycle can be performed sequentially, it is increasingly common that phases overlap or are iterative. In an overlapping relationship, the next phase of the product life cycle is initiated before the closing of the previous phase. The process groups are repeated within each phase of the project life cycle to guide the project to completion. This overlapping of process groups within phases can be seen in the *PMBOK® Guide*'s Figure 2-12.

In the *PMBOK® Guide*, three distinct project life cycles are discussed. You must know the differences between them and when they are generally preferred to be used.
- **Predictive**: used when a product is well understood (such as in building a house)
- **Iterative** or **incremental**: used when an organization needs to manage changing objectives and scope or when the partial delivery of a product is beneficial (such as with an environmental study that needs to be completed before plans can be finalized on a new airport runway)
- **Adaptive**: used when dealing with a rapidly changing environment and when requirements and scope are difficult to define in advance (such as with market driven software product development)

THE INFLUENCE CURVE

The influence curve demonstrates how important it is for organizations to plan projects. Note that the ability of a stakeholder to influence a change is high at the beginning of a project and decreases as the project progresses. Conversely, the impact or cost of a change is low at the beginning of a project and increases as the project progresses, as seen in Figure 2-8.

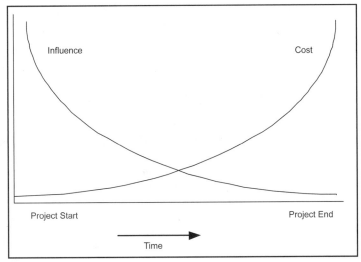

Figure 2-8
The Influence Curve

2

SAMPLE PMP EXAM QUESTIONS ON
PROJECT MANAGEMENT OVERVIEW

1. As the project manager on a software development project, your organization typically breaks up software projects into four smaller projects: requirements, design and development, quality assurance, and implementation. This is an example of:

   a) A work breakdown schedule
   b) A functional organization
   c) Progressive elaboration
   d) A project life cycle

2. One difference between a program and a portfolio is that:

   a) Programs are unique endeavors while portfolios are continuous
   b) A portfolio's success is dependent on the performance of the portfolio items, while a program's success is based on the program charter
   c) Programs are a set of related projects, while portfolios include all projects, related or not
   d) Programs have a business scope, while portfolios have a strategic scope

3. The life cycle that is preferred when the project scope is well-defined is called:

   a) Incremental
   b) Adaptive
   c) Predictive
   d) Iterative

Notes:

2

4. You have been working on your project in a foreign country. As you are exiting, you are forced to pay a substantial, "unofficial" exit fee. On your expense report, you:

a) Show nothing for the exit fee
b) Bury the exit fee in other parts of your expense report
c) Ask the sponsor how you could report the exit fee
d) Include the cost of the exit fee since you had to pay it to leave the country

5. Virtual teams rely on which of the following to be efficient and productive?

a) Collaborative online workspaces
b) Colocation of team members
c) Business partners who provide specialized expertise
d) Project management staff to handle administrative support

6. You are a project manager assigned to a project that is estimated to cost $3 million dollars. The product of the project is being developed because of speculation that a new market will be opening up. The senior management team has a high tolerance for risk and is willing to spend $3 million and potentially gain nothing in return. Embarking on this kind of project makes sense if the senior management team approves the project as part of an overall:

a) Tax reduction plan
b) Effort to induce investments
c) Portfolio strategy
d) Construction project

Notes:

7. You have joined a new organization that has a PMO. You have just been assigned to a project that appears to require specific technical resources used on many projects in the organization. What should you do first?

    a) Work with the functional manager to identify resources that may be backups
    b) Emphasize to your project sponsor the critical need for the special resources on this project
    c) Determine if the PMO has the authority to allocate resources to your project
    d) Contact the program manager to resolve resource constraints before they become a problem

8. As a project manager, you have been negotiating with a vendor on specific points of the business requirements. During the meeting between the vendor and the project sponsor, significant differences arise. Your responsibility includes:

    a) Restating the desires of the sponsor
    b) Explaining your understanding of the vendor's position
    c) Taking the side of the sponsor
    d) Sitting back and letting the vendor and sponsor work out their differences

Notes:

9. You are the project manager of the implementation of a new product line in a manufacturing facility in southeast Asia. The project is almost complete. The project has passed all quality control inspections except for one. All documented issues have been addressed, and many of the resources have been released. The project is slightly ahead of schedule but has a small budget overrun. The sponsor is onsite, and he has called a face-to-face meeting to get final signoff. He tells you not to worry about the missed quality item. What should you do first?

   a) Explain the significance of the missed quality item to the sponsor
   b) Follow the lead of the sponsor in communicating with the customer
   c) Communicate directly with the customer on the quality item
   d) Document the missed quality item in the issues log

10. On an information technology infrastructure project, one of architects is unhappy with the software development group. The architect starts to comment on how badly the developer did on the last project. What should you do?

   a) Take up the matter with the manager of the software development group and explain your team's frustration
   b) Ignore the architect's comments and give the architect and development group time for them to work out the issue
   c) Redirect the conversation to focus on solving the immediate problem
   d) Commiserate with the architect because you have had problems with that developer as well

Notes:

11. Life cycles are used to respond to high levels of change, and a great deal of stakeholder uncertainty relies on a set of requirements to be completed in very short iterations. These are called the product backlog and are reprioritized frequently. This life cycle is called agile, or:

   a) Adaptive
   b) Incremental
   c) Predictive
   d) Iterative

12. Any artifact, practice, or knowledge that can be used in your project and generally makes it easier to manage is called a(n):

   a) Organization process asset
   b) Infrastructure
   c) Commercial database
   d) Project management information system

13. Your consulting company has bid on an assignment to create an online course to help individuals prepare for the PMP exam. You have not created online course materials before, but you have just hired an expert in that field. How does this fit with the *PMI Code of Ethics and Professional Conduct*?

   a) It's not acceptable because you don't have experience in that field
   b) It's acceptable because you have other information technology experience and believe you can manage the expert
   c) It's acceptable because you have hired an expert who has the appropriate experience
   d) It's not acceptable because it's not consistent with your background, experience, or skills

Notes:

14. Interpersonal skills provide much of the foundation for building project management skills. However, managing a project requires additional competencies of:

a) Negotiating to acquire adequate resources
b) Motivating and inspiring team members
c) Effecting tradeoffs concerning project goals
d) Managing conflict among team members

15. A matrix organization that maintains many of the characteristics of a functional organization is called a:

a) Colocated organization
b) Weak matrix
c) Tight matrix
d) Projectized organization

Notes:

max technical training

513.322.8888
www.maxtrain.com

Get In. Get Out. Get Back to Work.

1.   A
2.   A
3.   C
4.   D
5.   A
6.   B
7.   B
8.   A
9.   D
10. C
11. D
12. D
13. C
14. D
15. B

2

ANSWERS AND REFERENCES FOR SAMPLE PMP EXAM QUESTIONS ON PROJECT MANAGEMENT OVERVIEW

Section numbers refer to the *PMBOK® Guide*.

1. **D Section 2.4 – Initiating**
   This could be the start of a WBS, but it's a typical software development life cycle.

2. **C Section 1.4 – Initiating**
   Know the differences between projects, programs, and portfolios.

3. **C Section 2.4.2.2 – Planning**
   Predictive life cycles include the traditional waterfall type used in information technology.

4. **A Section 2.3.1 – Executing**
   The cleanest way to handle the whole thing is to not include the exit fee because it may have the appearance of a bribe.

5. **A Section 2.3.1 – Executing**
   B) would not be a virtual team; C) and D) could be very helpful, but A) is more vital for virtual teams.

6. **C Section 1.4.2 – Initiating**
   Portfolio management is a strategic function.

7. **C Section 1.4.4 – Initiating**
   A), B), and D) may be actions to take after you have talked with the PMO.

8. **B Section 4.2.1 – Executing**
   Since you have been negotiating with the vendor, you may have more information, or have made other representations, that the sponsor is not aware of.

9. **A Code 5.3.1 – Closing**
   At some point, the project manager may have to do tasks in B), C), or D); however, the first step is to make sure the sponsor understands the quality issue.

2

10. C Section 3.2.4 & Section 4.2.2 – Initiating
First, you do not want to condone negative remarks that undermine another person's reputation. While A) and B) may be actions you will take, redirecting the conversation is the first thing you should do.

11. A Section 2.4.2.4 – Planning
This life cycle is also called agile, or change-driven.

12. A Section 2.1.4.1 – Initiating
B), C), and D) are all enterprise environmental factors.

13. B Section 2.2.2 – Initiating
Even though you don't personally have the experience to do the development, you have other information technology experience and have hired a professional. That meets the ethical standard for responsibility.

14. C Section 1.7 – Initiating
Negotiating, motivating, and managing conflict are examples of interpersonal skills.

15. B Section 2.1.3 – Initiating
Know the differences between a weak, balanced, and strong matrix.

CHAPTER 3 | **PROCESSES**

PROJECT MANAGEMENT PROCESSES, PROCESS GROUPS, AND THE INTERACTION OF PROCESSES

PMI has created a standard which documents the processes needed to manage a project. These processes are based on best practices that are practiced on most projects, most of the time. However, the *PMBOK® Guide* states that not all the processes need be, or even should be, applied to all projects all of the time. The project managers and their teams need to consider each process and determine if it is appropriate to their specific situation. This procedure is called tailoring by PMI. PMI feels that the processes and interactions among processes described in the *PMBOK® Guide* should serve as a standard for project management. Various methodologies and tools can be used to implement the framework for project management within an organization. Variances from the *PMBOK® Guide* standard are documented as part of the organization's **project management methodology**. The organization's project management methodology can, in turn, be tailored to fit the specific needs of the project based on customer requirements.

EXAM TIP

A diagram of the mapping of project management processes to the process groups and knowledge areas is found in the *PMBOK® Guide* Table 3-1.

Things to Know

1. **Project management processes** and **process groups**
2. **Work performance data**, **work performance information**, and **work performance reports**
3. Project management **process interactions**
4. The purpose of the **role delineation study**
5. Project management **performance domains**

Key Definitions

Input: a tangible item internal or external to the project that is required by a process for the process to produce its output.

Output: a deliverable, result, or service generated by the application of various tools or techniques within a process.

3

EXAM TIP
Be very familiar with
PMBOK® Guide Figures 3-1,
3-2, 3-3, 3-4, 3-5 and especially
Table 3-1.

Phase: one of a collection of logically related project activities usually resulting in the completion of one or more major deliverables. A project phase is a component of a project life cycle.

Process: a collection of related actions performed to achieve a predefined desired outcome. The *PMBOK® Guide* defines a set of 47 project management processes, each with various inputs, tools, techniques, and outputs. Processes can have predecessor or successor processes, so outputs from one process can be inputs to other processes. Each process belongs to one and only one of the five process groups and one and only one of the ten knowledge areas.

Process group: a logical grouping of a number of the 47 project management processes. There are five process groups, and all are required to occur at least once for every project. The process groups are performed in the same sequence each time: initiating, planning, executing, more planning and executing as required, and ending with closing. The monitoring and controlling process group is performed throughout the life of the project. Process groups can be repeated for each phase of the project life cycle. Process groups are not phases. Process groups are independent of the application area or the life cycle utilized by the project.

Tailoring: the act of carefully selecting processes and related inputs and outputs contained within the *PMBOK® Guide* to determine a subset of specific processes that will be included within a project's overall management approach.

Technique: a defined systematic series of steps applied by one or more individuals using one or more tools to achieve a product or result or to deliver a service.

Tool: a tangible item such as a checklist or template used in performing an activity to produce a product or result.

3

PROJECT MANAGEMENT PROCESSES AND PROCESS GROUPS

The *PMBOK® Guide* defines five process groups required for any project. They are:

- **Initiating**: defining and authorizing the project (or phase of the project)
- **Planning**: defining objectives, refining them, and planning the actions required to attain them
- **Executing**: integrating all resources to carry out the plan
- **Monitoring and Controlling**: measuring progress to identify variances and taking corrective action when necessary
- **Closing**: bringing the project or phase to an orderly end, including gaining formal acceptance of the result

The process groups are NOT project phases. In fact, it is not unusual to see all of the process groups represented within a single phase of a larger project.

In preparing for the PMP exam, take the time to read Chapter 3 of the *PMBOK® Guide* very carefully, along with Annex A1. PMI has put a lot of thought into the descriptions of the process groups, the interactions of the processes within them, and the relationships of each process group to the other process groups.

Each process group contains a number of processes, as listed below, but PMI has also identified ten topic-related groupings for the processes called **knowledge areas**. The processes associated with a particular knowledge area all address a single topic. For example, the processes within the time knowledge area address defining and planning the project schedule. The *PMBOK® Guide* is organized around these knowledge areas; Chapters 4 through 13 define each of the processes within a knowledge area in detail, covering the knowledge areas of integration, scope, time, cost, human resources, quality, communications, risk, procurement,

> **EXAM TIP**
> Know how tailoring applies to planning within a unique project setting.

3

and stakeholder management. This study guide is organized the same way to facilitate the exam candidate's study effectiveness.

The process group lists below show the process group, the processes in that group, and, in parentheses, the knowledge area in which that process is described.

Initiating Process Group

1. Develop Project Charter (Integration)
2. Identify Stakeholders (Project Stakeholder Management)

Planning Process Group

1. Develop Project Management Plan (Integration)
2. Plan Scope Management (Scope)
3. Collect Requirements (Scope)
4. Define Scope (Scope)
5. Create WBS (Scope)
6. Plan Schedule Management (Time)
7. Define Activities (Time)
8. Sequence Activities (Time)
9. Estimate Activity Resources (Time)
10. Estimate Activity Durations (Time)
11. Develop Schedule (Time)
12. Plan Cost Management (Cost)
13. Estimate Costs (Cost)
14. Determine Budget (Cost)
15. Plan Quality Management (Quality)
16. Plan Human Resource Management (Human Resources)
17. Plan Communications Management (Communications)
18. Plan Risk Management (Risk)
19. Identify Risks (Risk)
20. Perform Qualitative Risk Analysis (Risk)
21. Perform Quantitative Risk Analysis (Risk)
22. Plan Risk Responses (Risk)
23. Plan Procurement Management (Procurement)
24. Plan Stakeholder Management (Stakeholder Management)

Executing Process Group

1. Direct and Manage Project Work (Integration)
2. Perform Quality Assurance (Quality)
3. Acquire Project Team (Human Resources)
4. Develop Project Team (Human Resources)
5. Manage Project Team (Human Resources)
6. Manage Communications (Communications)
7. Conduct Procurements (Procurement)
8. Manage Stakeholder Engagement (Stakeholder Management)

Monitoring and Controlling Process Group

1. Monitor and Control Project Work (Integration)
2. Perform Integrated Change Control (Integration)
3. Validate Scope (Scope)
4. Control Scope (Scope)
5. Control Schedule (Time)
6. Control Costs (Cost)
7. Control Quality (Quality)
8. Control Communications (Communications)
9. Control Risks (Risk)
10. Control Procurements (Procurement)
11. Control Stakeholder Engagement (Stakeholder Management)

Closing Process Group

1. Close Project or Phase (Integration)
2. Close Procurements (Procurement)

WORK PERFORMANCE DATA, WORK PERFORMANCE INFORMATION, AND WORK PERFORMANCE REPORTS

In the *PMBOK® Guide*, a very specific distinction is made between work performance data, information, and reports.
- **Data** are raw observations and measurements that are identified as activities being performed
- **Information** is data that has been analyzed in context
- **Reports** are the physical or electronic representation of work performance information compiled in project documents

An easy way to remember these differences is that work performance data is typically an output of a process. Information is typically an input, while reports can be either an input or an output.

Work performance data is an output of the Direct and Manage Project Work process, while **work performance reports** are an output of the Monitor and Control Project Work process.

PROJECT MANAGEMENT PROCESS INTERACTIONS

Each process has **inputs**, **tools**, **techniques**, and **outputs** as defined in the *PMBOK® Guide*. You must take the time to learn the flow of the processes within each process group, the relations of processes across process groups, and how the outputs of one process become the inputs to other processes.

In order to help you understand how process groups flow and interact with one another, refer to Figure 3-1 on the following page.

3

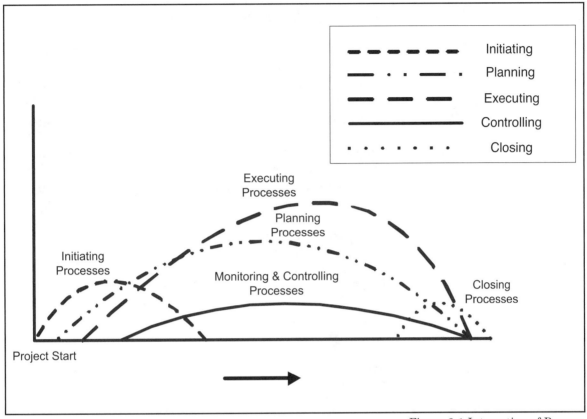

Figure 3-1 Interaction of Process Groups (*PMBOK® Guide* Figure 3-2)

ROLE DELINEATION STUDY

PMI conducts a role delineation study on a periodic basis to answer the question "what skills must project managers have to successfully lead and direct projects?" The results of the study provide the basis for the exam so that it mirrors the practices, methods, and systems of the project management profession. The study validates that the exam is measuring what project management professionals actually do on their jobs.

PERFORMANCE DOMAINS

Performance domains define the knowledge and skills needed by a project manager to complete the tasks in the project management processes competently. They are based on the role delineation study and described in the *PMI PMP Exam Content Outline—June 2015*. There are five performance domains, which match the five process groups described in the *PMBOK® Guide*. In addition, cross-cutting knowledge and skills have been defined for each domain, and these skills are highlighted in *Appendix X-3* in the *PMBOK® Guide*.

The domains are also used to specify the percentage of each type of question on the exam.

Performance Domains and Tasks

Figure 3-2 Performance Domains and Tasks

Figure 3-2 below is a summary of all performance domains, tasks, and cross-cutting skills a project manager needs to know. To see the full list, download the *PMI PMP Exam Content Outline—June 2015* from www.pmi.org.

Task	Domain
	I. Initiating
Task 1	Assess the feasibility of the project's product or service
Task 2	Define scope at a high level
Task 3	Align stakeholder expectations by conducting a stakeholder analysis
Task 4	Develop a high-level implementation approach
Task 5	Document scope, milestones, and deliverables in a project charter
Task 6	Document the project manager's authority and gain approval of the project charter
Task 7	Understand the expected business value and the project's alignment with organizational strategy by performing a benefits analysis
Task 8	Communicate and agreement of the approved project charter from all stakeholders
Cross-cutting skills	Business case and cost/benefit analysis, project selection criteria, techniques to identify risks and stakeholders, project charter components

Task	Domain
	II. Planning
Task 1	Establish project deliverables
Task 2	Create WBS and develop the scope management plan
Task 3	Develop the project budget and the cost management plan
Task 4	Develop the project schedule and the resource management plan
Task 5	Develop the human resources management plan
Task 6	Develop the communications management plan
Task 7	Develop the procurement management plan
Task 8	Develop the quality management plan
Task 9	Develop a change management plan
Task 10	Create a risk register and develop a risk management plan
Task 11	Get approval for the project management plan from stakeholders
Task 12	Hold a project kick-off meeting
Task 13	Create a stakeholder register and develop a stakeholder management plan
Cross-cutting skills	Organizational charts, tools, and techniques to gather requirements, create WBS, estimate time and cost, plan resources, sequence the flow of tasks, planning elements and purposes, process improvement approaches, efficiency principles, environmental impact assessments, contract types, and communication methods
	III. Executing
Task 1	Get internal and outsourced resources and manage resources
Task 2	Execute project tasks and lead team
Task 3	Use quality management plan for quality assurance
Task 4	Execute on approved changes and corrective actions
Task 5	Manage risks by using the identified response plans to minimize negative and maximize positive risks
Task 6	Keep stakeholders engaged and informed
Task 7	Manage stakeholder involvement, support, and expectations
Cross-cutting skills	Tools and techniques for monitoring, budgeting, and continuous improvement, knowledge of scope, quality tools, interaction of tasks, and vendor management

3

Task	Domain
	IV. Monitoring and Controlling
Task 1	Measure performance, note variances, identify corrective action(s)
Task 2	Manage project changes to deliver the stated business needs
Task 3	Use quality standards to confirm that deliverables meet customer requirements
Task 4	Maintain the risk register as project changes or situations occur
Task 5	Use the issue log to document corrective action(s) and track potential project changes
Task 6	Continuously improve project management processes
Task 7	Verify procurement activities comply with project objectives
Cross-cutting skills	Tools, techniques, and metrics to analyze and measure performance, knowledge of thresholds, variance limits, integrated change control, financial principles, risk management techniques, quality verification, and validation techniques
	V. Closing
Task 1	Get sponsor or customer acceptance of project deliverables
Task 2	Transition deliverables to assigned owners
Task 3	Get financial, legal, and administrative closure
Task 4	Communicate final project information to all stakeholders
Task 5	Review documented lessons learned from throughout the project life cycle
Task 6	Archive project deliverables
Task 7	Obtain feedback from customer on satisfaction
Cross-cutting skills	Knowledge of compliance, contract and financial closing requirements, review, feedback, archiving techniques, and transition plans

Exercise 3-1

In the WAV files connected with this study guide, there are charts of the 47 processes with the inputs, tools, techniques, and outputs for each process. Print out each chart, in color if possible, and cut out and rearrange the processes for each one of the process groups.

HINT: See *PMBOK® Guide* Figures 4-1, 5-1, 6-1, 7-1, 8-1, 9-1, 10-1, 11-1, 12-1, and 13-1.

The charts of each process will also make a very portable quick reference you can use as a study aid.

SAMPLE PMP EXAM QUESTIONS ON MANAGEMENT PROCESSES

1. Performing the initiating processes at the start of each phase helps focus the project on:

    a) Potential risks that should be identified
    b) Additional requirements for the project
    c) The business need for the project
    d) Procurement items that may be needed in the phase

2. What is likely to take place during the executing processes?

    a) Changing resource availability, updating activity durations, and rebaselining
    b) Controlling changes and recommending corrective or preventive actions
    c) Verifying that defined processes are complete and concluding project activities
    d) Refining the project objectives and defining the steps to take during the project

3. What defines the planning process group actions?

    a) Completes the work defined in the project management plan to satisfy the project specifications
    b) Defines a new project or new phase by obtaining authorization to start the project or phase
    c) Finalizes all activities across all process groups to formally close the project or phase
    d) Establishes the project scope, refines the objectives, and defines the course of action required to achieve the objectives

Notes:

4. Which definition defines the executing process group?

   a) Completes the work defined in the project management plan to satisfy the project specifications
   b) Defines a new project or new phase by obtaining authorization to start the project or phase
   c) Finalizes all activities across all process groups to formally close the project or phase
   d) Establishes the project scope, refines the objectives, and defines the course of action required to achieve the objectives

5. You are a new project manager for a startup software company. As you began your career there, you notice that the project management methodology is very informal. One of the first things you should do is:

   a) Educate the organization on why a more formal methodology is needed
   b) Establish a PMO in the organization
   c) Nothing; if the process is working, an informal methodology could certainly be used
   d) Communicate to the project team that they need to follow your proven methodology

6. The project manager is always responsible for:

   a) Ensuring that the project is highly profitable
   b) Hiring and firing members of the project team
   c) Selecting projects that can be accommodated
   d) Determining what processes are appropriate

7. Aligning the stakeholders' expectations with the project's purpose, giving them visibility about the scope and objectives, and showing how their participation will help the project is a key purpose of which process group?

   a) Planning
   b) Initiating
   c) Executing
   d) Monitoring and controlling

Notes:

8. Evaluating the impact of change, developing a root cause analysis, and performance tracking are skills used in the _____ performance domain.

   a) Closing
   b) Monitoring and controlling
   c) Planning
   d) Executing

9. When a project is divided into phases, how do the process groups apply?

   a) It's usually not necessary to go through the initiating process group again
   b) Planning of a prior phase will probably be sufficient for the next phase
   c) Monitoring and controlling will be simplified because of your experience in a prior phase
   d) All process groups may be repeated for each phase

10. A set of interrelated actions performed to create a product, service, or result and assure the effective flow of the project throughout its life cycle is:

   a) Product management process
   b) Project management process
   c) Project life cycle
   d) Project management plan

11. Which of the following statements is true regarding project management?

   a) It should focus on the life cycle of product-oriented processes
   b) Project requirements may be met without intentionally managing interrelated activities
   c) It is an integrative undertaking that requires alignment among processes
   d) Projects exist within a closed system in the organization

Notes:

12. What is the key benefit obtained from project planning processes?

    a) Allows the project manager to manage stakeholders' expectations and coordinate project resources
    b) Provides a way to track and review progress and manage changes aggressively
    c) Aligns stakeholders' expectations with the project's purpose
    d) Helps the team define the strategy, tactics, and course of action to complete a project

13. Performance domains together with their _____, _____, and _____ define the project management profession.

    a) Tasks, knowledge, skills
    b) Schedules, costs, scope
    c) Performance reports, control charts, skills
    d) Organization charts, domain impact, project plan

14. Significant changes occurring throughout the project life cycle can most likely trigger:

    a) A rewrite of the project charter and a need for new authorization
    b) The need to revisit one or more of the initiating or planning processes
    c) Assignment of a new project manager and team members
    d) Cancellation of the project and contracts

Notes:

15. On your project, the quality assurance department has delivered the results of a recent system test. The results have been analyzed by several team members and compiled into an overall analysis of the project. The team then summarizes the analysis for presentation to the project sponsor. This summary is called:

a) Work performance completed
b) Work performance data
c) Work performance information
d) Work performance reports

Notes:

3

ANSWERS AND REFERENCES FOR SAMPLE PMP EXAM QUESTIONS ON MANAGEMENT PROCESSES

Section numbers refer to the *PMBOK® Guide*.

1. **C Section 3.3 – Initiating**
   A), B), and D) are all planning processes.

2. **A Section 3.5 – Executing**
   B) take place in monitoring and controlling processes, C) take place in closing processes, and D) take place in planning processes.

3. **D Section 3.4 – Planning**
   A) defines the executing process group, B) defines the initiating process group, and C) defines the closing process group.

4. **A Section 3.5 – Executing**
   B) defines the initiating process group, C) defines the closing process group, and D) defines the planning process group.

5. **C Section 3.0 – Initiating**
   A project management methodology does not have to be formal or mature. The objective of any project is to satisfy the needs of the customer.

6. **D Section 3.0 – Initiating**
   A) projects don't have to be profitable; B) project managers may not have the authority to hire and fire; C) project managers don't always get to choose the projects they work on.

7. **B Section 3.3 – Initiating**
   Some of this information is captured in the stakeholder register.

8. **B Section IV – Monitoring and Controlling**
   These activities show up in several different knowledge areas.

9. **D Section 3.1 – Initiating**
A) some sort of initiating process is preferred at the start of each phase; B) and C) prior phases are likely to be very different and potentially use different resources, so the planning and monitoring and controlling are not likely to be the same or easier.

10. **B Section 3.0 – Initiating**
The goal of project management is to integrate both product and project processes.

11. **C Section 3.0 – Initiating**
A) product management processes don't reflect how projects are managed; B) may be true, but it will be accidental; D) projects don't exist in a closed environment, they require input from the organization and beyond.

12. **D Section 3.4 – Planning**
A) is a benefit of the executing processes, B) is a benefit of the monitoring and controlling processes, and C) is a benefit of the initiating processes.

13. **A Section Overview – Initiating**
The performance domains give an overview of what skills project managers must have to be successful in managing projects.

14. **B Section 3.4 – Monitoring and Controlling**
Although A), C), and D) may happen, the most likely action is that the project manager should revisit various planning processes to determine what may need to change, if anything.

15. **D Section 3.8 – Executing**
Know the difference between data and information; data is typically an input to many processes, and information is an output.

INTEGRATION

CHAPTER 4 | **INTEGRATION**

INTEGRATION MANAGEMENT

You need to know that the PMP exam addresses critical project management functions that ensure coordination of the various elements of the project. The *PMBOK® Guide* explains that the processes in project management are integrative in nature. They involve making tradeoffs among competing objectives to meet stakeholders' needs and expectations. Integration processes drive the associated knowledge area processes within each of the process groups, and all process groups are addressed by one or more integration management processes. These processes interact with each other as well as with the other nine knowledge areas.

It is important to note that integration occurs within as well as outside a project. For example, project scope and product scope must be integrated, and project work must be integrated with the ongoing other work of the organization (such as operations and deliverables from various technical specialties). One of the key tools or techniques used to integrate the processes and measure project performance is **earned value management** (EVM). **Work performance reports** are introduced in this chapter and utilized for the **performance reviews** and **performance reporting** mentioned in chapters on Time Management (Chapter 6), Cost Management (Chapter 7), and Communications Management (Chapter 10).

Interactions often have a domino effect that can be overt or subtle. Interactions engender tradeoffs, and the project manager must be able to orchestrate these interactions as the project flows and changes throughout its life cycle. For example, increased resource usage in one area may adversely affect the schedule in another area. Complicating the picture is the fact that the project manager has to answer to many stakeholders and must constantly communicate upward, downward, and laterally to ensure success. The integration management knowledge area of the exam stresses how the different knowledge areas interact to continuously improve our ability to plan, perform, and predict work.

EXAM TIP

Although integration management is the first knowledge area to be discussed here (following the sequence of the *PMBOK® Guide*), we recommend that you revisit it after all other knowledge areas have been reviewed.

Integration management occurs throughout the **project life cycle**, from project start to close. The project manager faces many challenges that differ from those faced by functional or operational managers. The project manager must coordinate the integration of:
- Project work with ongoing operations
- Product scope and project scope
- Schedule, budget, metrics, and reporting
- Skills, knowledge, and deliverables from vendors, stakeholders, and the performing organization
- Risks and staffing plans
- Performance and quality objectives

All six of the project integration management processes contain the tools and techniques of **expert judgment**. **Facilitation techniques**, **meetings**, **analytical techniques**, and the **project management information system** (PMIS) are also common among many of the integration processes. These tools and techniques must be defined, clearly understood, and used in developing the project charter and other key documents of the project.

Things to Know

1. The six processes of integration management:
   - **Develop Project Charter**
   - **Develop Project Management Plan**
   - **Direct and Manage Project Work**
   - **Monitor and Control Project Work**
   - **Perform Integrated Change Control**
   - **Close Project or Phase**
2. The purpose of a **business case**
3. The **project statement of work**
4. The benefits of good **facilitation techniques**
5. The importance of **expert judgment**
6. The contents of the **project charter**
7. The contents of the **project management plan**
8. **Tailoring** and project management
9. How to run **meetings** well
10. The purpose of a **project management information system**
11. Types of **analytical techniques**

12. **Forecasting methods**
13. How to manage **change requests**
14. **Configuration management** and the **change control system**
14. Key **interpersonal skills** for success

Key Definitions

Application area: a category of projects that share components that may not be present in other categories of projects. For example, approaches to information technology projects are different from those for residential development projects, so each is a different application area.

Change control: the procedures used to identify, document, approve (or reject), and control changes to the project baselines.

Change management: the process for managing change in the project. A change management plan should be incorporated into the project management plan.

Enterprise environmental factors: external or internal factors that can influence a project's success. These factors include controllable factors such as the tools used in managing projects within the organization and uncontrollable factors that have to be considered by the project manager such as market conditions or corporate culture.

Expert judgment: judgment based on expertise appropriate to the activity. It may be provided by any group or person, either within the organization or external to it.

Organizational process assets: any formal or informal processes, plans, policies, procedures, guidelines, and on-going or historical project information such as lessons learned, measurement data, project files, and estimates versus actuals.

4

Progressive elaboration: the progressive improvement of a plan as more specific and detailed information becomes available during the course of the project.

Project management information system: the collection of tools, methodologies, techniques, standards, and resources used to manage a project. These may be formal systems and strategies determined by the organization or informal methods utilized by project managers.

Project management methodology: any structured approach used to guide the project team through the project life cycle. This methodology may utilize forms, templates, and procedures standard to the organization.

DEVELOP PROJECT CHARTER PROCESS

This process is used to formally authorize a new project or validate an existing project for continuation into the next phase. Projects are initiated as a result of a problem to be solved, an opportunity, or a business requirement. Therefore, many organizations often have more projects than resources to complete the projects. A method to select high priority projects must be used. In one method, projects are selected based on the business need and cost benefit analysis that are contained in the business case.

Business Case

An organization often has many identified projects; usually, more projects exist than available resources can handle. Therefore, it is a good practice to follow a project selection and prioritization process in which justification and cost benefit analyses of the project are evaluated. The business case is the document that justifies why a project should be accomplished. In some cases, the customer or requesting organization writes the business case. Projects come about based on business needs such as:

EXAM TIP

In order to successfully initiate a project, the project manager must understand the expected business value of the project by evaluating the business case or performing a benefits analysis.

EXAM TIP

Know the reasons business cases are created. See the *PMBOK® Guide* section 4.1.1.2.

- **Market demand**: for the organization to stay in business, it must produce competitive products
- **Customer request**: the sponsoring organization needs and will to pay for new functionality
- **Legal or social requirements**: Sarbanes-Oxley requirements and reports for accounting practices

It is important to review the business case at the end of each phase to ensure that the business need for the project still exists.

Project Statement of Work

An input to the Develop Project Charter process is the statement of work (SOW). The SOW is a narrative description of the products or services to be delivered by the project. Since the products or services can be delivered by either an internal or an external organization, the SOW is also an important artifact of the project procurement management knowledge area. The *PMBOK® Guide* calls a SOW for external organizations a contract statement of work. A SOW generally includes:

- The **business need** or reasons the product or service is required
- The description of the **product scope** or **product requirements** with as much detail as possible to support project planning and estimates
- The **strategic plan** and how the project supports the organization's strategic goals to facilitate the process of deciding which projects to select

> **EXAM TIP**
>
> In order for a project to be initiated, a benefit analysis should have been conducted with relevant stakeholders. This ensures that the project is aligned with organizational strategy and expected business value.

Facilitation Techniques

To facilitate means to make easier. The project manager is regularly presented with issues, challenges, problems, and conflict within the project setting and must have the ability to work through various situations to ensure the project progresses. It is the project manager's role to make it easier for the team to succeed. Some facilitation techniques that are typically employed are brainstorming, conflict resolution, problem solving, and meeting management.

4

Expert Judgment

Expert judgment is a tool and technique used in many processes throughout the *PMBOK® Guide*. Developing a project charter requires that expert judgment be used in potentially several ways. Expertise may be leveraged in understanding a problem to be solved and also in defining an approach to solve it. Past experiences with similar projects can also aide in outlining high-level scope and identifying risks, constraints, and assumptions for a project.

EXAM TIP

The project manager must be able to articulate the expected business value of the project in order to successfully deliver on stakeholder expectations. Reviewing the approved project charter with stakeholders is a great way to ensure alignment.

The Project Charter

The project charter is an important document that establishes a project. It typically:
*   Contains the **business need**, purpose, and justification for the project
*   Includes the **product requirements**
*   Identifies the assigned project manager and the project manager's **authority** level for the project
*   Could be the signed **contract** or **agreement** when procurement processes are used
*   Defines the **goals** and **objectives** of the project (sets the project direction)
*   Is approved by key stakeholders
*   Shows organizational, environmental, and external **constraints** and **assumptions**
*   Contains the summary **budget** and milestone schedule

The *PMBOK® Guide* highlights the iterative nature of projects and the use of **progressive elaboration** in the development of projects. The Develop Project Charter process demonstrates that when projects are initiated, not all information is available in a detailed form. The objective of this process is to get the project initiated and to assign resources to explore the detailed needs of the project.

The **project charter** might be considered a "1st draft" in understanding project requirements. The project charter is an input to the planning processes of the Develop Project Management Plan, Collect Requirements, and Define Scope processes to help define the project management plan, requirements, and scope statement.

Some aspects critical to the measurement of the success of a project are outlined in the project charter. These key aspects are:
- Purpose and **project objectives**
- Project sponsor or **authority**
- Project description, requirements, and boundaries
- **Success criteria**
- **Acceptance criteria**
- **Identified risks**
- Initial **work breakdown structure** (WBS), preliminary milestones, and summary budget
- **Project manager assignment** (with responsibility and authority level)

CASE STUDY: THE LAWRENCE GARAGE PROJECT

This case study will be used throughout this study guide. You will be asked to prepare various documents relating to the case study as you go through the initiating, planning, executing, and monitoring and controlling processes. Since this study guide is organized following the presentation of topics in the *PMBOK® Guide*, the exercises are not necessarily in the order in which they would be completed in a real project. Although this case study is not on the exam, it is provided here to strengthen your understanding and application of project management disciplines and techniques.

Examples of each of the documents you need to create for a project are provided for you to compare with your results. There are no right or wrong documents as you will undoubtedly develop different levels of detail in your answers. The examples provided simply demonstrate our interpretation of one approach. Any of the currently available project management software packages will provide adequate results for these case study exercises.

Case Study Overview

The goal of the Lawrence garage project is to build a garage on the property of Mr. and Mrs. Lawrence. They have been parking their RV and their son's boat in their backyard in a fenced, gated area. However, their city has passed a new ordinance that prohibits residents with RVs, boats, off-road vehicles, or utility trailers from storing or parking such vehicles on their property or on the street for more than 72 hours. These vehicles must now be parked or stored in a "building suitable for that purpose." Mr. and Mrs. Lawrence have decided to build a garage.

They have hired an architectural firm to develop the plans and specifications for the garage. These have been completed, along with all the working drawings and a bill of materials, and the architect has also agreed to be Mr. and Mrs. Lawrence's project manager, help them select a contractor, and work with the contractor's project manager.

Acme Construction and Engineering (ACE) has won the bid to build the garage. You have been appointed to be the project manager for ACE. You were not involved in the bidding process.

The garage is designed to be big enough to hold a motor home or travel trailer and a boat. It is 50 feet long and 30 feet wide with two 14-feet-high by 12-feet-wide roll-up doors for the RV and boat and a standard 3-foot side door. There are two windows on the side of the garage that has the standard door, three windows on the side with no door, and two windows in the back wall. Inside there are to be lights and electrical outlets to be used when working on the RV or boat. The plans also show RV hookups inside (water, electric, and sewer) and there should be a drain in the floor. There should be a small bathroom (commode and sink only) in the back corner of the garage and a utility sink adjacent to the bathroom.

The exterior walls are to be stucco, and the roof is to be asphalt shingles to match the existing house. Walls should be of 2X6 construction to allow for greater insulation.

Since there is an existing home, utilities are already present. The architect has submitted the plans to the city, but these plans have not yet been approved and a permit has yet to be issued.

Case Study Exercise

Exercise 4-1: Create a project charter for the Lawrence garage project. It really should be written by the general manager for the contractor, but she has asked you to write a draft and then the two of you will polish it together.

DEVELOP PROJECT MANAGEMENT PLAN PROCESS

The Develop Project Management Plan process integrates all the **subsidiary plans** from the various knowledge areas into one cohesive whole. This complete, consistent, and coherent document is the project management plan. It is crucial that the project manager and the project team spend sufficient time in creating the project management plan because this document serves to reduce project uncertainty, improve the efficiency of work, provide a better understanding of project objectives, and provide a basis for monitoring and controlling. This key document also serves as a communication and educational tool for stakeholders on the project.

4

Developing a project management plan is a crucial step in planning a project. The *PMBOK® Guide* emphasizes four planning concepts:

- Planning begins during the starting phase of a project
- Planning does not end until the project is finished
- Planning is an intellectual process that runs through all the other processes of the project
- Planning is iterative; as the project proceeds, the project team must plan, replan, and plan again

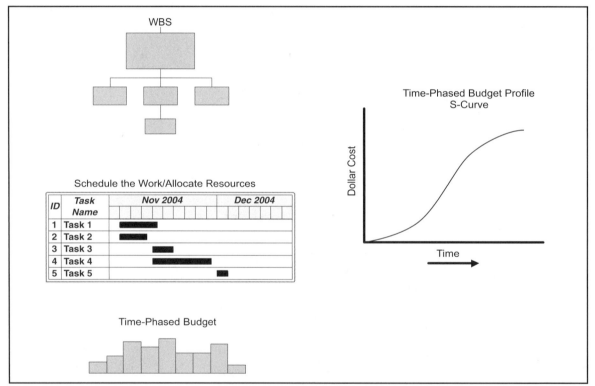

Figure 4-1
Planning Outputs

In planning a project, first the project objectives are clearly defined and a WBS is created to define the work in more detail; this WBS is used to schedule the work with resources assigned to obtain a time-phased budget. This time-phased budget is used to obtain the planned value S-curve. All of these can been seen in Figure 4-1 above.

The Project Management Plan

A project management plan is more than just a set of instructions. It is not just a WBS or an output of a scheduling tool. The objective of developing a project management plan is to eliminate crisis by preventing anything from falling through the cracks. Creating it is an **iterative process**, and it is used to guide project execution. It is a collection of documents that identify the project assumptions and decisions. The project management plan provides both a standard communication tool throughout the lifetime of the project and a verification and agreement tool on objectives and requirements from the stakeholders. The project management plan is documented and approved by both the customer and the sponsor. All baselines for tracking, control, analysis, communications, and integration elements are incorporated into the project management plan.

The key elements of the project management plan may ultimately include:
- The selected **project life cycle** to be used
- The project management approach or strategy to be used in the definition of the selected processes and tools and techniques
- How a standard process could be tailored
- How project work will be managed and executed
- How changes will be monitored and controlled
- How **configuration management** will be performed
- How the integrity of performance baselines will be maintained, including **requirements traceability**
- How the project will be closed
- The description of other phases or related projects in multiphase or multiprogram environments
- The level of frequency of communication and techniques for communicating with stakeholders
- The scope, schedule, and cost **baselines** and how each will be maintained.
- The schedule or milestone chart of key management reviews for progress, issue resolution, and decisions

EXAM TIP

Know the key elements of the project management plan.

4

4

> **EXAM TIP**
>
> The project management plan is begun in the Develop Project Management Plan process, but it gets updated by other planning, executing, and monitoring and controlling processes.

- **Subsidiary management plans**, including the following plans and their definitions:
  - **Scope management plan**: provides guidance on how project scope is defined, documented, verified, managed, and controlled; it is created before performing the six processes of scope management
  - **Requirements management plan**: tracks how requirements are managed, analyzed, documented, and prioritized and traces changes to requirements throughout the project life cycle
  - **Schedule management plan**: describes how changes to the schedule will be managed
  - **Cost management plan**: describes how cost variances will be managed based on the needs of the stakeholders
  - **Quality management plan** and **process improvement plan**: describe how the project management team will implement its quality policy
  - **Human resource plan**: describes how and when human resources will be brought into and taken off the project team
  - **Communications management plan**: addresses the collection, distribution, access to, and updates of project information
  - **Risk management plan**: documents procedures to manage risk throughout the project
  - **Procurement management plan**: describes how procurement processes will be managed
  - **Stakeholder management plan**: provides guidance on how the various stakeholders can be best involved in the project
  - Subsidiary management plans may also include a **change management plan** and/or a **configuration management plan**

Tailoring

Tailoring is a form of **expert judgement** that is used to accommodate unique situations. Since no two projects are identical, and each project brings with it new challenges and risks, the project manager must consider how standard or unique a project is and determine whether or not tailoring is needed.

It can be expected that the organizational process assets we use for a project will be available to be used, re-used, or modified to accommodate the uniqueness of a project. This is a form of tailoring.

A project manager may also have significant experience in a particular type of project, such as software implementations, but each project has different stakeholders with different needs; therefore, there could be many unique attributes to a project which need to be considered. A project manager must make tailoring decisions and balance the project constraints as well as meet the acceptance criteria of the deliverables.

DIRECT AND MANAGE PROJECT WORK PROCESS

Project execution consists of the primary processes whereby the project management plan is put into action. Most of the project resources and costs are expended in this process. Important inputs to this process are the project management plan, the **enterprise environmental factors**, and the **organizational process assets** because these inputs influence how the project team carries out its work.

In order for the project management plan to be executed, work is assigned to resources by the project manager and then monitored for the completion of deliverables and work results. Therefore, the key outputs of this process are **deliverables**, **change requests**, and **work performance data**. Since project activities are performed by people, the project manager must understand the organizational structure and individual motivations of the people working on the project. These skills are addressed within the Develop Project Team process of the human resource management knowledge area (Chapter 9).

Meetings

No project is successful without appropriate and well-run meetings. Meetings are initiated to provide information exchanges, discussion opportunities for brainstorming, or decision making forums.

Good meetings tend to have the following characteristics:
- They are prepared for in advance
- A specific agenda, purpose, objective, and timeframe is provided to participants
- They include individuals that are appropriate to the topic being discussed to facilitate the meeting's purpose

Project Management Information System

The project management information system (PMIS) is a system that can include both manual and automated systems used to gather, integrate, and disseminate the outputs of the Develop Project Management Plan process. It is used by the project management team to create and control changes to the project management plan. The PMIS can contain many subsystems, such as project management scheduling software, information/progress reporting systems, web links to related systems, and a configuration management system.

MONITOR AND CONTROL PROJECT WORK PROCESS

As the work of a project is carried out, deliverables or work results are created. Issues arise or changes come about in the form of change requests. The project manager and team continuously monitor performance by comparing actual performance against the project management plan and determining if corrective or preventive actions are necessary. **Schedule forecasts,**

cost forecasts, and **work performance information** are key inputs to this process. In some cases, defects may be found which may require rework. As a result of any of the above issues, one or more change requests are created. These change requests will impact project deliverables and may increase project costs and schedules. In some cases, a mid-project evaluation may be needed.

Mid-project evaluations are conducted while project work is still in progress. The main purpose of such evaluations is to determine if objectives are still relevant and if these objectives are being met. **Lessons learned** should also be documented at this time instead of waiting to document them until the project has been completed.

EXAM TIP

Throughout the project life cycle, the project manager should engage in lessons learned activities to identify areas for improvement.

A third party or people outside the team should be used to conduct mid-project evaluations so that results are less biased. Mid-project evaluations, however, often cause stress and conflict within the project team for the following reasons:
- Evaluations can be disruptive, take time away from project work, require review meetings to be scheduled, and result in significant changes
- Evaluations can generate the impression in project team members that they are not to be trusted (i.e., that "Big Brother is watching")
- Evaluations can be arbitrary or inconsistent, making evaluation results misleading, skewed, or invalid
- Management may misuse evaluation results to identify and punish poor performance
- Management may use evaluation results to support personal agendas or goals
- The identification of significant problems often results in drastic changes
- The project could be terminated

However, mid-project evaluations can prove extremely beneficial if conducted well. Such evaluations may involve project participants who enter or leave a project at phase transitions. Those departing can provide useful

4

insights based on their intimate experience with the project. Those arriving could provide "fresh eyes" and new methods for achieving project objectives. An enterprise quality assurance function, a corporate audit department, or a project office with proven processes and experienced staff may help mid-project evaluations be more readily accepted.

Analytical Techniques

The Monitor and Control Project Work process may use a variety of analytical techniques in order to forecast potential outcomes. Your organization may or may not use these. It is important to know that many techniques are available, and you should be familiar with the definitions of each:

- Regression analysis
- Grouping methods
- Causal analysis
- Root cause analysis
- Forecasting methods
- Failure mode and effect analysis
- Fault tree analysis
- Reserve analysis
- Trend analysis
- Earned value management
- Variance analysis

> **EXAM TIP**
> Read and get to know the definitions of each of these analytical techniques.

Forecasting Methods

As part of reporting project performance, project managers must understand and provide estimates of future performance of the project in terms of a forecast.

General categories of forecasting methods include:
- **Time series**: the use of historical data as the basis of a forecast; an example of a time series method of forecasting is the use of a cost performance index (CPI) in earned value analysis
- **Economic method**: the identification of a causal effect on the outcome of the project; for example, the volatility of the price of oil may have a

significant impact on the project team's ability to control material costs

- **Judgmental method**: the use of expert judgment as a basis for costs forecasts

There are other methods that can be used as well. The project manager must determine the appropriate method and apply it to performance reports. Forecasting techniques are described in more detail in Chapter 7 (Cost).

Managing Change Requests

In the latest edition of the *PMBOK® Guide*, PMI consolidated several change request topics into the overall concept of managing change requests. Change requests are an output of the Monitor and Control Project Work process. They can come in several forms, such as corrective action, preventative action, or defect repair. In other cases, change requests are additions to the agreed-upon scope and should, once documented and approved, involve adjustments to the project baseline.

> **EXAM TIP**
>
> Know the differences between corrective action, preventive action, and defect repair as each relates to change requests.

PERFORM INTEGRATED CHANGE CONTROL PROCESS

Changes in projects are inevitable. The Perform Integrated Change Control process coordinates changes across the entire project by determining that a change has occurred, managing the change when it does occur, and ensuring that changes are controlled and agreed upon. Through the use of a change control system, the project manager is able to be in control and make necessary adjustments to ensure a project's success.

The Perform Integrated Change Control process works hand-in-hand with the Control Communications process in communications management to integrate the subsidiary change control processes found in the scope, time, cost, quality, risk, and procurement knowledge areas.

Within the Perform Integrated Change Control process there are tools and techniques called **change control tools**. These tools should facilitate configuration and change management by managing the change requests and resulting decisions. In change control meetings, members of the **change control board** (CCB) follow the change control process to ensure that requested changes to the project and product scope are properly considered and documented. Both the **configuration management** and **change control** procedures are integrated with the overall PMIS.

Configuration Management

Configuration management is a systematic procedure for managing change. Configuration management protects both the customer from unauthorized changes by project staff and the project staff from new or undocumented requirements changes from the customer. A configuration management system:
- Is a collection of formal documented procedures used to apply technical and administrative direction and control
- Is a subsystem of the PMIS
- Includes the processes that define how project deliverables and documents are controlled, changed, and approved
- In many areas includes the change control system

The purpose of configuration management is to ensure compliance to stated requirements by:
- Identifying and documenting the characteristics of the configurable items
- Accounting for and managing any changes to the configurable items
- Verifying and auditing the integrity and consistency of the configurable items as they have been defined

Change Control System

The change control system is also a collection of formal documented procedures that define how project change requests are submitted, validated, recorded, approved or rejected, communicated, and worked within the project. In many areas the change control system is a subset of the configuration management system.

CLOSE PROJECT OR PHASE PROCESS

The Close Project or Phase process is one of the two processes in the closing process group, the other being the Close Procurements process described in the project procurement management chapter (Chapter 12). There are many questions about the similarities and differences between these two processes. The Close Project or Phase process includes the step-by-step activities needed to conclude a project or phase, such as:

- Transitioning the project's product, service, or results to production or the next phase as needed to satisfy the exit criteria of the project or phase
- Updating the **organizational process assets** updates

Know that the Close Procurements process supports the Close Project or Phase process because it also verifies the acceptance of project procurement deliverables. In addition, a project manager will use many tools, such as expert judgement, analytical techniques, and meetings to bring a project to successful closure.

KEY INTERPERSONAL SKILLS FOR SUCCESS

The interpersonal skill highlighted in this chapter is:

Leadership

Leadership involves getting project work completed by helping others focus on achieving the objectives of the project. Key elements of leadership include respect and trust. Respect is addressed as part of the PMI *Code of Ethics and Professional Conduct*.

At the beginning of a project, it's critical that project managers create a vision that is compelling, motivating, and inspiring. Throughout a project, the project manager will need to help team members grow into a high performing team by building and maintaining trust, mentoring, coaching, and providing feedback.

As a project closes, leadership skills are necessary to gain acceptance of the project deliverables and ensure that the project objectives are met and that stakeholders are satisfied.

SAMPLE PMP EXAM QUESTIONS ON INTEGRATION MANAGEMENT

1. Documenting the approved configuration and the status of changes that are approved and implemented is accomplished by:

   a) Configuration status accounting
   b) Configuration verification and audit
   c) Configuration control
   d) Configuration identification

2. What is an action to make sure that work performed in the future aligns with the project management plan?

   a) Plan risk responses
   b) Take corrective action
   c) Develop preventive actions
   d) Repair defects

3. You are preparing a schedule for your project. You have had a discussion with Juan, the information technology manager, about when his business systems analyst will be needed on the project. He agrees that the timing will work. In preparing the project schedule, you note the date the business systems analyst will join the project. You have:

   a) Identified an assumption
   b) Identified a constraint
   c) Set up a milestone to be met
   d) Set up a task on your critical path

4. What kinds of actions are taking place during the Direct and Manage Project Work process?

   a) Collecting stakeholder requirements and developing a WBS
   b) Identifying quality criteria and the risks related to them
   c) Obtaining materials and equipment and managing team members
   d) Deciding on information to be shared and how it will be distributed

5. Controlling includes:

    a) Measuring and assessing performance and distributing reports
    b) Managing resources and generating work performance data
    c) Preparing and coordinating subsidiary management plans
    d) Determining corrective action, replanning, and following up

6. It seems that projects in your organization never finalize documentation of the results. You want to avoid this problem on your current project. What could you implement?

    a) Establish a closing procedure that includes stakeholders' comments
    b) More clearly define the exit criteria for the acceptance of scope
    c) Verify that deliverables meet the quality standards
    d) Validate that completed project deliverables meet scope requirements

7. Your latest project involves a new use for your existing product. When you reviewed the business justification for the project, you discovered that your marketing department had identified demand for this new use and that your competitors were developing something similar. The new use will allow a region to access its banking system more easily, which will improve the economy. These requirements should be documented in the:

    a) Business case
    b) Procurement statement of work
    c) Project scope statement
    d) Project management plan

Notes:

8. The project manager reviews work results of completed project scope activities and measures these against the:

   a) Requirements
   b) Project management plan
   c) Scope management plan
   d) Work breakdown structure

9. You are reviewing the status of your project and want to make sure that change requests consider the effect on the project as a whole. Which of the following reports are likely to be needed?

   a) Project charter and statement of work
   b) Resource availability, schedule, and cost data
   c) Documentation of verified configuration items
   d) Lists of preventive and corrective actions

10. You have taken over a project to update your corporate website. It is about 30% complete, and you know there is a project management plan available. It appears that there are technical problems with the links on the website. What is the best way to address this situation?

    a) Bring up the issue at the regular team meeting
    b) Set up a separate meeting to brainstorm and address the issue
    c) Put the issue on the agenda of the meeting with the approvers of that project phase
    d) Post the issue on the shared website so that everyone will have access to the discussion

11. You are managing your website redesign project and want to take into account the organization's environment. Which factors should you consider?

    a) Communication requirements of the organization
    b) The organization's work authorization system
    c) Processes to review time and expense reports
    d) Procedures to define issues

Notes:

12. In multi-phase projects, the Close Project or Phase process can be applied to:

    a) Completing make-or-buy analyses and decisions
    b) The project scope and activities applicable to the project phase
    c) Only the final phase of the project
    d) Only one project phase

13. Once initiated, a project should be halted if:

    a) The project manager leaves
    b) The statement of work by the buyer is incomplete or ambiguous
    c) The project is found to be supportive of the organization's strategic goals
    d) The business need no longer exists

14. When may partial deliverables be considered accepted as part of closing a project?

    a) When the schedule is in danger of being overrun
    b) When there are approved change requests
    c) When a project is cancelled
    d) When the project is not finished

15. Each week, your project team provides you with schedule progress, the extent to which quality standards are being met, and resource utilization details. These are all examples of:

    a) Earned value analysis
    b) Work performance information
    c) Monte Carlo simulation
    d) Quality audits

Notes:

4

ANSWERS AND REFERENCES FOR SAMPLE PMP EXAM QUESTIONS ON INTEGRATION MANAGEMENT

Section numbers refer to the *PMBOK®* *Guide*.

1. **A Section 4.5 – Monitoring and Controlling**
   B) includes actions to ensure the correctness of the configuration and that changes have been documented; C) is the overall process; D) is selecting the configurations items to track and maintaining accountability for changes.

2. **C Section 4.3 – Executing**
   Know the differences between corrective action, preventive action, and defect repair.

3. **A Section 4.0 – Initiating**
   Assumptions are factors that, for planning purposes, are considered to be true, real or certain. See the *PMBOK®* *Guide* Glossary.

4. **C Section 4.3 – Executing**
   A), B), and D) are all planning activities, not executing activities.

5. **D Section 4.4 – Monitoring and Controlling**
   A) is monitoring; B) is directing and managing project work; C) is planning.

6. **A Section 4.6 – Closing**
   All of these are good practices, but the goal is to improve documentation.

7. **A Section 4.1.1.2 – Initiating**
   The business case may also include a cost/benefit analysis.

8. **B Section 4.4 – Monitoring and Controlling**
   The Monitoring and Controlling Project Work process is concerned with comparing actual project performance against the project management plan.

9. **B** **Section 4.5.1.3 – Monitoring and Controlling**
A) is possible, but B) is a better answer; C) is a configuration audit; D) gives potential results of change requests.

10. **B** **Section 4.3.2.3 – Executing**
Meetings should have an agenda, and it's not very effective to mix meetings that have different purposes, e.g., exchanging information, brainstorming options, or making decisions. Posting doesn't actively address the issue.

11. **B** **Section 4.4.1.6 – Monitoring and Controlling**
A), C), and D) are all organizational process assets.

12. **B** **Section 4.6 – Closing**
A) the Close Project or Phase process (an integration process) is more broad than finalizing the make-or-buy decision in the Plan Procurement Management process; C) the Close Project or Phase process can occur at multiple points during a project; D) The Close Project or Phase process should occur for every project phase.

13. **D** **Section 4.4 – Monitoring and Controlling**
The Monitor and Control Project Work process includes the assessment of whether or not a project will meet the objectives it was initiated for.

14. **C** **Section 4.6.1.2 – Closing**
When a project phase ends, you might also accept partial project deliverables.

15. **B** **Section 4.3.3.1 – Executing**
A) is a tool and technique of the Control Costs process; C) is a tool and technique of the Develop Schedule and Perform Quantitative Risk Analysis processes; D) are structured, independent reviews of compliance with project policies, processes, and procedures.

CASE STUDY SUGGESTED SOLUTION

Exercise 4-1
Project Charter for the Lawrence Garage Project

This project's goal is to build a recreational vehicle (RV) and boat garage on the property of Mr. and Mrs. David Lawrence in Anytown, United States. The garage will be built according to the architectural plans and specifications provided by the owners.

The major deliverables are the garage structure, including a finished interior, and a driveway connecting to the existing driveway per the plans and specs.

The garage is being built to comply with new city codes that require boats and/or RVs stored for more than 72 hours to be enclosed in a structure suitable for that purpose.

This project is scheduled to start on or about May 20, 2014 and should be completed within 120 calendar days. The initial budget is $37,500.

The project sponsors are Mr. and Mrs. Lawrence. Their project manager is Scott Hiyamoto, who is also the architect. The project manager for Acme Construction and Engineering (ACE) is [Your Name Here]. Mr. Hiyamoto is acting as the agent of the Lawrences and has full authority to make decisions on their behalf. Mr. and Mrs. Lawrence will direct all their communications through Mr. Hiyamoto. [Your Name Here] is acting as the agent for ACE and has full authority to make decisions on its behalf. All employees of ACE and all of its subcontractors, along with their employees, will direct all their communications through [Your Name Here].

Signed

_____   _____
Susan Ruzicka, General Manager   David E. Lawrence
Acme Construction & Engineering   Property Owner

_____   _____
[Your Name Here]   Scott Hiyamoto, Managing Partner
Acme Construction & Engineering   Hiyamoto, Kwame, & Blum, LLP

SCOPE

CHAPTER 5 | **SCOPE**

5

SCOPE MANAGEMENT

Project scope management questions on the PMP exam cover diverse yet fundamental project management topics. Defining and managing the scope or the amount of work of the project is an important aspect of project management. Before the project manager jumps into executing the project, there must be an understanding of how scope will be defined, developed, monitored, controlled, and verified. This understanding is defined within the **scope management plan**. Additionally, the **requirements management plan** describes how requirements will be analyzed, documented, and managed.

These plans are important because there must be an understanding of how the project scope is broken down into smaller, more manageable components and how these components are controlled. The requirements management plan, the updated and refined scope statement, work breakdown structures (WBSs), scope validation, and scope changes are the topics covered.

The role of the project manager includes defining the work, making sure that only the work of the project is being completed and preventing additional work (scope creep) not defined in the project. The *PMBOK® Guide* does not advocate "gold plating" or giving the customer more than what was asked for.

In the project context, the term "scope" may refer to:
- **Product scope**, which consists of the features and functions of a product or service with results measured against the product requirements
- **Project scope**, which describes the work that must be done to deliver a product, service, or result with completion being measured against the project management plan

Project scope management is concerned with defining and controlling both what is and what is not included in the project.

> **EXAM TIP**
>
> The *PMBOK® Guide* places enormous emphasis on the tool or technique of decomposition, which is breaking down the major project deliverable into smaller, more manageable elements. These elements are further subdivided until the deliverables are specific enough to support the planning, executing, monitoring and controlling, and closing process group activities. The resulting collection of elements that define the total scope of the project is called the work breakdown structure.

5

Things to Know

1. The six processes of scope management:
   - **Plan Scope Management**
   - **Collect Requirements**
   - **Define Scope**
   - **Create WBS**
   - **Validate Scope**
   - **Control Scope**
2. The importance of the **scope management plan**, **requirements management plan**, and **requirements traceability matrix**
3. What **product analysis** is
4. The purpose and contents of the **project scope statement**
5. All aspects of the scope baseline (the **scope statement**, the **WBS** and the **WBS dictionary**)
6. The **code of accounts** (and how it differs from the **chart of accounts**)
7. That a **WBS dictionary** is
8. The purpose of **control accounts**
9. The differences between the **Validate Scope** and **Control Quality** processes
10. Controlling **scope changes**
11. Key **interpersonal skills** for success

Key Definitions

100% Rule: the WBS should represent the total work at the lowest levels and should roll up to the higher levels so that nothing is left out, and no extra work is planned to be performed.

Chart of accounts: the financial numbering system used to monitor project costs by category. It is usually related to an organization's general ledger.

Code of accounts: the numbering system for providing unique identifiers for all items in the WBS. It is hierarchical and can go to multiple levels, each lower level containing a more detailed description of a project deliverable. The WBS contains clusters of elements that are child items related to a single parent element; for example, parent item 1.1 contains child items 1.1.1, 1.1.2, and 1.1.3.

Control account: the management control point at which integration of scope, budget, and schedule takes place and at which performance is measured.

Decomposition: the process of breaking down a project deliverable into smaller, more manageable components. In the Create WBS process, the results of decomposition are deliverables, whereas in the Define Activities process, project deliverables are further broken down into schedule activities.

Planning package: a component of the work breakdown structure that is below the control account to support known uncertainty in project deliverables. Planning packages will include information on a deliverable but without any of the details associated with schedule activities.

Requirements traceability matrix: a matrix for recording each requirement and tracking its attributes and changes throughout the project life cycle to provide a structure for changes to product scope. Projects are undertaken to produce a product, service, or result that meets the requirements of the sponsor, customer, and other stakeholders. These requirements are collected and refined through interviews, focus groups, surveys, and other techniques. Requirements may also be changed through the project's configuration management activities.

Rolling wave planning: a progressive elaboration technique that addresses uncertainty in detailing all future work for a project. Near-term work is planned to an appropriate level of detail; however, longer term deliverables are identified at a high level and decomposed as the project progresses.

Scope baseline: the approved detailed project scope statement along with the WBS and WBS dictionary.

Scope creep: the uncontrolled expansion of a product or project scope without adjustments to time, cost, and resources.

WBS dictionary: houses the details associated with the work packages and control accounts. The level of detail needed will be defined by the project team.

Work breakdown structure (WBS): a framework for defining project work into smaller, more manageable pieces, it defines the total scope of the project using descending levels of detail.

Work package: the lowest level of a WBS; cost estimates are made at this level.

PLAN SCOPE MANAGEMENT PROCESS

The Plan Scope Management process is the first of many planning processes that detail out the "how" of the **project management plan**. This process defines how the scope will be defined, validated, and controlled.

Scope Management Plan

The scope management plan defines the processes that the project team will follow for:
  - Preparing the detailed scope statement
  - Creating the WBS
  - Establishing how the WBS will be maintained and approved
  - Specifying how formal acceptance will be obtained
  - Managing how requests for changes will be processed

The scope management plan helps the project team avoid scope creep.

Requirements Management Plan

The requirements management plan is an important document that is used with the requirements documentation to assist the project manager in communication, thereby lessening possible future issues of managing the requirements' baseline, project changes, conflict issues, and project risks. The requirements management plan:

- Describes how requirements will be managed
- Describes how changes to requirements will be identified and classified
- Describes how changes to requirements will be integrated, tracked, reported, and approved
- Defines the requirements prioritization process
- Discusses metrics to measure the product
- Details the attributes to be tracked in the requirements traceability matrix

COLLECT REQUIREMENTS PROCESS

This process defines the activities needed to determine and define the features and functions that meet the needs and expectations of the sponsor, customer, and other stakeholders. It uses the **project charter**, **scope management plan**, **requirements management plan**, and **stakeholder management plan** as inputs and many tools and techniques to elicit and document stakeholder needs into **product requirements** (for example, product specifications, performance, quality, security, etc.) or **project requirements** (business guidelines, project management practices, delivery schedules, etc.).

As individual requirements are gathered, they are combined into the requirements documents and reviewed with stakeholders. The requirements documentation must meet the business need for the project. Once the requirements documentation has been reviewed and approved as complete, consistent, traceable, and clear, it is baselined as part of the scope baseline.

5

Requirements Traceability Matrix

A requirements traceability matrix is an output of the Collect Requirements process and is critical to controlling scope creep in that it looks at each requirement of the project and links those requirements directly back to a specific project objective. Managing requirements tightly ensures that the project will stay focused on the delivery of the project objectives. Those that do not will most likely experience excessive change requests and a high potential for **scope creep**.

DEFINE SCOPE PROCESS

Although a high level description of the project and product is prepared in the Develop Project Charter and Plan Scope Management processes, it is refined here with additional information about the project and the product, service, or result of the project.

Product Analysis

EXAM TIP
It is important to know the differences between the project charter and the project scope statement. For example, the project charter provides the project purpose or justification; however the project scope statement defines the acceptance criteria necessary to validate the success of the project. See Table 5-1 in the *PMBOK®* *Guide*.

Managing a project that affects or creates a product has different considerations than those projects that produce a service or result. Considerations of material purchases, marketing, and integration with customer environments are a few. For example, a project that has been initiated to enhance the design of a riding lawnmower product line must consider the full bill of materials and product design to determine the specific changes that are necessary to the bill of materials to ensure the product produced will meet market expectations.

Project Scope Statement

The project scope statement is the basis for future project decisions and is critical to project success. It provides the basis for agreement on project scope between the project team and the customer. The scope statement contains enough information to allow stakeholders to document their agreement on the:

- Project's **scope description**, which elaborates on the characteristics of the product, service, or result
- **Deliverables**
- Project **exclusions**
- **Constraints and assumptions**
- **Acceptance criteria**

A detailed project scope statement:
- Provides documentation for future project decisions
- Can have multiple levels as project work goes through decomposition
- Is refined or revised to reflect approved changes to the requirements of the project
- Is referred to in some organizations as the scope of work or statement of work (SOW)

In order to ensure project success, project **objectives** must be defined. Good objectives should be clear, defined well, and feasible. Project objectives should follow the SMART guideline. That is, objectives must be:
- **S**pecific — clear with no ambiguity
- **M**easurable — with quantifiable indicators of success
- **A**ssignable — with responsibility resting on an individual or organization
- **R**ealistic — achievable within the constraints
- **T**imely — with specific duration and due dates

Case Study Exercise

Exercise 5-1: Once the project charter from Exercise 4-1 has been accepted by the stakeholders, you need to develop a scope statement for the Lawrence Garage Project. It should be detailed enough so that all stakeholders know what is being done, but since there are architectural drawings already existing, you will not need to describe every board.

5

CREATE WBS PROCESS

In order to further understand and define the work of the project, it must be broken down into smaller, more manageable components. The resulting deliverable-oriented hierarchical structure is known as the **work breakdown structure (WBS)**. The lowest levels of the WBS are the planned units of work the project team must execute in order to achieve the project objectives and results.

Scope Baseline

The scope baseline is the key output of the **Create WBS** process. It consists of three distinct components—the scope statement, the WBS, and the WBS dictionary.

Work Breakdown Structure

The scope baseline, which includes the WBS, provides input to other planning processes, such as the Define Activities, Estimate Cost, Determine Budget, Identify Risks, and Perform Qualitative Risk Analysis processes.

Any work not defined in the WBS is outside the scope of the project; therefore the WBS is the foundation of the project and contains the building blocks of project work. The WBS:
- Breaks the project into smaller pieces, those at the lowest level being known as **work packages**
- Defines the total scope of the project using descending levels of detail
- Is deliverable oriented
- Contains items that are assigned unique identifiers (i.e., **code of accounts**)

The benefits of using a WBS are that it:
- Facilitates communication and a common understanding of project scope
- Provides a framework for project identification
- Brings focus to project objectives
- Forces a breakdown of project work into smaller work packages that are more easily assigned and tracked

> **EXAM TIP**
> Know the difference between a planning package and a work package.

- Creates work packages that are small enough for more accurate estimates
- Identifies holes or weak areas of project scope requirements
- Facilitates performance measurement
- Clearly defines responsibilities
- Facilitates progress status reporting, problem analysis, and the tracking of time, cost, and performance
- Allows for improved handling of change control requests
- Is available for reuse with appropriate modifications for similar projects

The following are some guidelines for developing a WBS:
- Utilize the project team for help
- Identify higher levels before breaking down into more detailed levels
- Know that some components will break down into more detailed levels than others
- Work down toward tangible deliverables (work packages), keeping in mind that:
  - The effort to produce the deliverable can be confidently estimated
  - The types of skills required for the deliverable can be evaluated
  - Required resources can be determined
  - Costs can be determined and confidently estimated
  - The deliverable can be easily tracked

Some examples of names of WBS levels are:
- **Program:** a grouping of programs and projects all linked to a common objective or goal
- **Project:** the summation of all work to be completed to achieve a unique product, service, or result
- **Control Account:** a summary level of the WBS used to categorize and measure progress

- **Work Package:** the lowest-level item of the WBS. Cost estimates are made at this level; it is sometimes stipulated that a work package is not more that 80 hours of effort. Work packages assist in risk identification and may be broken down further during the Define Activities process into schedule activities and even smaller tasks
- **Planning Package:** a known set of work that will need to be performed without the specific details of schedule activities
- **Schedule Activity:** a further subdivision of the work package
- **Task:** work not necessarily listed in the WBS that is the lowest level of effort on the project

Figure 5-1
The Control Account

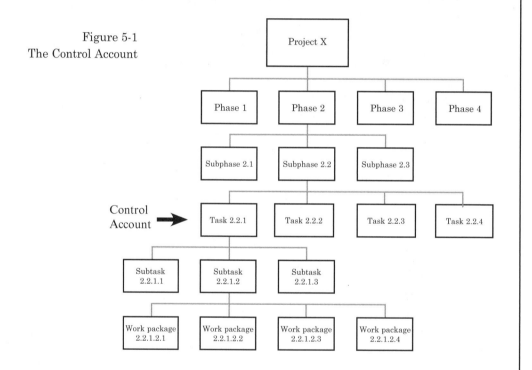

WBS Dictionary

This document provides detailed information about each component of the WBS. The value of the WBS dictionary is that it includes everything needed to perform and validate the work defined in the WBS, such as resources required and acceptance criteria for each work package defined.

The outputs of the Create WBS process are the WBS, of course, but also the WBS dictionary and the project scope statement. These three items are combined and called the **scope baseline** which is a component of the project management plan.

Control Accounts

An important aspect of the WBS are control accounts which are management control points at which the integration of scope, schedule, and cost takes place and at which performance is measured. Management control of projects at every work package level can be very time consuming, so the use of selected management points of the WBS is recommended. Therefore, a control account could control one or more work packages. The advantages of control accounts are that:
- All **earned value** performance measurement in a project should take place at the control level
- They are the building blocks of **performance measurement**
- The sum of the control accounts will add up to the total project value

There is no recommended standard for the appropriate dollar size of a control account. Some organizations use a rule of thumb, such as 300 hours—but the size should be whatever is manageable.

In the example in Figure 5-1 on the prior page, the control account is placed at the section level (2.2.1, 2.2.2, etc.) and individual or small group efforts at the sub-task or work package levels are rolled up to the control account plan to be managed by the project manager and reported to stakeholders. This avoids micromanaging individual team members and smoothes out minor variances from the plan.

Case Study Exercise

Exercise 5-2: After the scope statement from Exercise 5-1 has been approved, develop the WBS to show the breakdown of the work that needs to be done to deliver the complete Lawrence Garage Project.

VALIDATE SCOPE PROCESS

The Validate Scope process is the process of getting formal acceptance of completed project deliverables from **stakeholders** who may be the project sponsor, clients, or customers. The Validate Scope process confirms work has been completed correctly and satisfactorily because it:

- Involves the review of work results by conducting audits, reviews, and inspections
- Is similar to the Control Quality process as both involve checking work products
- Is different from the Control Quality process because it focuses on the acceptance criteria of the deliverables (i.e., the scope of work) instead of the validation of the deliverables
- Is often performed after the Control Quality process, but the two processes can be performed in parallel
- Determines whether work results conform to requirements
- Documents the formal acceptance and signoff of deliverables
- Could generate change requests for defect repairs
- Can be repeated at the end of each project phase

Validation is typically determined through a predefined process of **inspection** that should have been agreed upon as part of the **scope statement** and **requirements management plan**.

Validate Scope versus Control Quality

At the end of every project, the project manager and the project team will go through an exercise of obtaining customer acknowledgement that the project has delivered upon its objectives. The **acceptance criteria** which were developed during the development of the project charter, scope statement, and project plan will be used in determining the final outcome of the project—success or failure.

EXAM TIP

Know the difference between accepted deliverables and validated deliverables, as well as the process of which they are outputs.

The process for validating scope is very different from the process that will be discussed in Chapter 8, the Control Quality process. The Control Quality process is focused on the lower levels of the WBS and ensures that each component of the project is being delivered according to the **detailed requirements** or **specifications**. Theoretically, as each level of deliverables is validated as part of the Control Quality process, the likelihood of a successful scope verification will increase, but will not be guaranteed.

CONTROL SCOPE PROCESS

The Control Scope process monitors the status of both project and product requirements. It also manages changes to the requirements and the **scope baseline**. Frequently, a **scope change** also requires adjustments to project objectives, cost, time, and quality. Therefore, a scope change usually means updates to the **project management plan**, which includes the scope baseline and other baselines defined. Project documents and organizational process assets may also be updated. In most cases, replanning is necessary. Change requests and work performance information are other outputs of the Control Scope process.

5

Controlling Scope Changes

A key responsibility of the project manager is to ensure that an appropriate **change management** procedure be in place for the project. A change management process is implemented for the sole purpose of reviewing and determining whether or not requested changes are considered within the scope or outside the scope of the project. Typically a change management process includes an entity called a **change control board** (CCB) whose responsibility it is to review and either approve or deny change requests.

A CCB should include membership of the project team and other key stakeholders, including the customer. The CCB's purpose is to objectively evaluate the requests being submitted and determine an action to be taken.

A strong change control process reduces the risk of scope creep.

KEY INTERPERSONAL SKILLS FOR SUCCESS

The interpersonal skill highlighted in this chapter is:

Decision Making

Many decisions must be made to finalize requirements and scope. Decision styles that may be used by a project manager:

- Command
- Consultation
- Consensus
- Coin flip

These styles may be affected by the relationships among stakeholders, including trust and acceptance, or external constraints such as time and quality. There are a number of decision-making models, all of which include some form of these steps:

1. Define the problem
2. Generate ideas about solutions
3. Evaluate solutions
4. Communicate and gain support for the solution selected
5. Conduct a post-implementation evaluation of the result

SAMPLE PMP EXAM QUESTIONS ON SCOPE MANAGEMENT

1. You are the project manager for a business process improvement project for a strategic business process in which several new business process needs have been identified as planning packages in the WBS. As part of the initial activities for the project, a sub-team has been assigned the responsibility to detail out the deliverable expectations of these new planning packages. This process is known as:

   a) Analogous estimating
   b) Bottom-up estimating
   c) Discovery
   d) Decomposition

2. You are the project manager on a project critical to your organization. Controlling the overall project scope is determined to be a key measurement of success. In order to ensure success, you should:

   a) Spend time detailing out the project scope and work that is excluded
   b) Bring on experts to ensure that the best possible product is delivered
   c) Communicate the project's priority to everyone to assure adequate resources
   d) Over budget for resources so that you are assured of staying on schedule

3. You have taken over a project that started 2 months ago. It is critical to your organization. Management created a project charter and wants you to get started immediately because of a major corporate announcement that will occur in a few months. The prior project manager put together a plan describing how scope will be defined, validated, and controlled. What should you do first?

   a) Develop a scope management plan
   b) Define the scope
   c) Analyze the stakeholders
   d) Collect requirements

Notes:

4. The Validate Scope process outputs include:

   a) Influencing the factors that create project scope changes
   b) Documenting the completed deliverables that have been accepted
   c) Obtaining the stakeholders' formal acceptance of the completed project scope
   d) Monitoring specific project results to determine whether they comply with relevant quality standards

5. Which of the following organizational process assets are likely to influence the Plan Scope Management process?

   a) Issue and detect identification procedures
   b) Infrastructure and personnel administration
   c) Organization culture and market conditions
   d) Policies and lessons learned

6. In order to enhance planning and managing work on the project:

   a) Produce a WBS at the highest level
   b) Use lower levels of decomposition in a WBS
   c) Implement a work authorization system
   d) Make planning meetings open to everyone who has an interest

7. You are a project manager on a medical device development project. The organization you work for has strict operational guidelines that must be followed in order to meet FDA regulations with regard to changes to the project deliverable scope. These guidelines are an example of:

   a) Enterprise environmental factors
   b) Organizational process assets
   c) Organizational systems
   d) Regulatory filing requirements analysis

Notes:

8. The formal and informal policies, procedures, and guidelines that can impact how a project is managed are called:

   a) Communication requirements analysis
   b) Organizational systems
   c) Organizational process assets
   d) Enterprise environmental factors

9. The process of defining and documenting stakeholders' needs to meet the project objectives is called:

   a) Plan Scope Management
   b) Plan Stakeholder Management
   c) Collect Requirements
   d) Define Scope

10. You are asked by your sponsor to describe how the formal verification and acceptance of the project deliverables will be obtained. The document that should include this information is the:

    a) Executive sponsor sign-off
    b) Procurement statement of work
    c) Scope management plan
    d) Communications management plan

11. You have been assigned to a project that is in trouble. After talking with the customer you find that there were over 50 requests for scope changes, and the customer is frustrated that the previous project manager only implemented 5 changes that weren't of importance to the project's success. Upon further review, you determine that no formal change control process was implemented and therefore the customer's expectation had been that the 50 changes were going to be completed. This is an example of:

    a) Chaos
    b) Scope creep
    c) A poor performing project manager
    d) An out-of-control customer

Notes:

12. Your project is nearing completion and you have scheduled a deliverable review meeting with your customer for next week. The objective of the meeting will be to verify that each project deliverable has been completed satisfactorily and has been accepted by the customer. This is an example of the _____ process.

    a) Validate Scope
    b) Scope Identification
    c) Inspection
    d) Management by Objectives

13. You are managing a project that is on schedule with sixty work packages completed. One day a stakeholder tells you that she has found a critical product defect that had not been discovered when the initial project plan and charter were developed. To solve this problem you should:

    a) Do nothing because it wasn't part of the initial scope
    b) Submit a change request to the change control board and re-baseline the project if and when the change is approved
    c) Communicate the error to senior management
    d) Update the project schedule and budget to reflect the necessary changes

14. A key input to the Define Scope process is:

    a) Requirements documentation
    b) WBS dictionary
    c) Project scope statement
    d) Risk register

Notes:

5

15. A good scope change control system would define:

a) The procedures by which the project scope and product scope can be changed
b) When and if any changes can be made
c) The procedures by which ONLY the product scope can be changed
d) How to more completely detail the project scope to eliminate the risk of any scope changes

Notes:

ANSWERS AND REFERENCES FOR SAMPLE PMP EXAM QUESTIONS ON SCOPE MANAGEMENT

Section numbers refer to the *PMBOK® Guide.*

1. **D** **Section 5.4 – Planning**
   A Planning package has known work content but has not been decomposed into detailed schedule activities.

2. **A** **Section 5.3 – Planning**
   The best defense is a great offense, and projects are no exception. The *PMBOK® Guide* advocates that the more time spent planning and working through the details of the project early on, the less likely there will be any surprises later on.

3. **D** **Section 5.2 – Planning**
   The prior project manager has completed the scope management plan. The next step is to collect the needs and expectations of the stakeholders.

4. **B** **Section 5.5 – Monitoring and Controlling**
   A) is the purpose of Control Scope; C) is the purpose of Validate Scope; and D) is the purpose of Control Quality.

5. **D** **Section 5.1.1.4 – Planning**
   A) occurs during executing, monitoring, and controlling; B) and C) are enterprise environmental factors.

6. **B** **Section 5.4.2 – Planning**
   A) A high-level WBS does not give you sufficient detail in order to manage work; C) A work authorization system will not solve your issue unless you have sufficient detail defined; D) Open meetings may enhance planning and communication, but they do not necessarily enhance your ability to manage the work.

7. **B** **Section 2.1.5 – Planning**
   A) Enterprise environmental factors may include items such as an organization's culture, human resources, and marketplace conditions; C) An

organizational system defines whether or not an organization is project-based or non-project based; D) Summarizes the informational needs of the project stakeholders.

8. C **Section 2.1.4 – Planning**
A) Summarizes the informational needs of the project stakeholders; B) An organizational system defines whether or not an organization is project-based or non-project based; D) Enterprise environmental factors may include items such as an organization's culture, human resources, and marketplace conditions.

9. C **Section 5.2 – Planning**
Stakeholder input into requirements, conditions, and capabilities is key to a project's success.

10. C **Section 5.1.3.1 – Planning**
The scope management plan identifies the verification and acceptance criteria for the project.

11. B **Section 5.6 – Monitoring and Controlling**
Scope creep is defined as uncontrolled changes.

12. A **Section 5.5 – Monitoring and Controlling**
B) Scope identification is a planning process; C) inspection is the tool and technique used within the Validate Scope process; D) management by objectives is a system of managerial leadership.

13. B **Section 5.6 – Monitoring and Controlling**
Change request are outputs of the Control Scope process

14. A **Section 5.3.1 – Planning**

15. A **Section 5.6 – Monitoring and Controlling**
Changes are inevitable. Having a good change control process that addresses both project and product scope will ensure manageability of the process.

CASE STUDY SUGGESTED SOLUTIONS

Exercise 5-1
Scope Statement for the Lawrence Garage Project

As stated in the project charter, the goal of this project is to build a recreational vehicle (RV) and boat garage on the property of Mr. and Mrs. David Lawrence in order to comply with a new city ordinance regarding RV and boat parking.

The garage is designed to be big enough to hold a motor home or travel trailer and a boat, with room at the back for miscellaneous storage and a small workshop area. The garage will be 50 feet long and 30 feet wide with 2 14-foot-high by 12-foot-wide roll-up doors for the RV and boat and a standard 3-foot side door. There are 2 windows on the side of the garage that has the standard door, 3 windows on the side with no door and 2 windows in the back wall. Wall height is 16 ft.

The roof is to use engineered trusses, manufactured by an approved supplier. Inside there are to be lights and electrical outlets to be used when working on the RV or boat. The interior is to be heated and air conditioned. The plans also show RV hookups inside (water, electric, and sewer), and there should also be a drain in the floor. There should be a small bathroom (commode and sink only) in the back corner of the garage and a utility sink adjacent to the bathroom.

The exterior walls are to be stucco and the roof is to be asphalt shingles to match the existing house. Walls should be of 2X6 construction to allow for greater insulation.

Since there is an existing home, utilities are already present. The garage will have its own electric service and its own hot water heater. The architect has submitted the plans to the city, but they have not yet been approved, and permits have not yet been issued.

The major deliverables are the garage structure, including finished interior, and a driveway connecting to the existing driveway per the site plan. No landscaping is included in the specifications.

This project is scheduled to start on or about May 20, 2014 and complete within 120 calendar days. The initial budget is $37,500.

Assumptions: the plans will be approved by the city as submitted. The plans will be approved by June 1, 2014. The plans and specifications, which are attached by reference, are complete and sufficient to plan, execute, and complete this project. Anything not included in the plans and specifications is explicitly outside the scope of this project.

Constraints: construction permits for residential units, including detached garages, state that construction must be complete within 185 days from the issue date. The owner states that any change order resulting in an increase of $1,000 from the baseline cost, or when accumulated change orders exceed $2,000, a formal review of the proposed change or changes must be held with the owners, architect, and general contractor all present.

Mr. and Mrs. Lawrence have stated that they would be extremely happy with the finished project as long as their current boat and 20-foot-long motor home can be accommodated with room for the Lawrences to easily walk around with both the boat and the RV parked in the garage. Mr. Lawrence would also like to comfortably accommodate repair work from within the garage that may be necessary from time to time.

Attached by reference:

Site plan
Foundation and slab plan
Framing plan
Electrical plan
Plumbing plan

Heating, ventilating, and air conditioning plan
Door and window specifications
Exterior finish plan
Interior finish plan
Roof truss engineering specifications
Driveway specifications
Bill of materials

Exercise 5-2
WBS List for the Lawrence Garage Project

WBS	Task Name
0	Lawrence recreational vehicle garage
1	Project management
1.1	Finalize plans and develop estimates
1.2	Sign contract and notice to proceed
1.3	Apply for permits
1.4	Execute and control project
1.4.1	Track progress
1.4.2	Weekly status meeting
2	Site work and foundation
2.1	Clear and grub lot
2.2	Install temporary power service
2.3	Install underground utilities
2.4	Excavate for foundations
2.5	Install forms and rebar
2.6	Pour concrete for foundation and slab
2.7	Cure concrete
2.8	Strip forms
2.9	Perform foundation/slab inspection
2.99	Foundation complete
3	Framing
3.1	Install mudsills
3.2	Frame walls
3.3	Frame corners
3.4	Install roof trusses
3.5	Complete roof framing
3.6	Conduct framing inspection
3.99	Framing complete
4	Exterior
4.1	Install wall sheathing
4.2	Install roof decking
4.3	Install felt, flashing, and shingles
4.4	Hang exterior doors
4.5	Install windows
4.6	Wrap exterior
4.7	Apply brown coat
4.8	Cure brown coat 7 days
4.9	Apply finish coat

5

WBS List for the Lawrence Garage Project Continued

4.99	Exterior complete
5	Interior
5.1	Finish rough-in plumbing
5.2	Rough-in electrical
5.3	Rough-in HVAC
5.4	Rough-in communication (phone, cable, computer, alarm)
5.5	Utility inspections
5.6	Place wall insulation
5.7	Place ceiling insulation
5.8	Install drywall on walls
5.9	Install drywall on ceilings
5.10	Tape and float drywall
5.11	Prime all walls and ceilings
5.12	Paint all walls and ceilings
5.13	Install bath and storage cabinets
5.14	Complete plumbing
5.15	Complete electrical
5.16	Complete communications (phone, cable, computer, alarm)
5.17	Complete HVAC
5.99	Interior complete
6	Grounds work
6.1	Grade for driveway and sidewalk
6.2	Setup driveway and sidewalk forms
6.3	Lay out driveway rebar
6.4	Pour concrete driveway and sidewalks
6.5	Cure concrete
6.6	Remove forms
6.99	Grounds complete
7	Final acceptance
7.1	Complete final inspection for certificate of occupancy
7.2	Cleanup for occupancy
7.3	Perform final walk-through inspection
7.4	Complete punch list items
7.99	Project complete

TIME

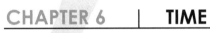

CHAPTER 6 | TIME

6

6

TIME MANAGEMENT

Time management is that portion of project management concerned with the project schedule. It includes defining the project activities, ordering the activities in their logical sequence, estimating the effort and duration of each activity, and building an overall project schedule. Time management also includes managing the schedule once the project is under way, since the actual amount of time it takes to complete activities does not always match the estimates.

The PMP exam focuses time questions heavily around the precedence diagramming method (PDM), the critical path method (CPM), the critical chain method, and three-point estimating. You must be familiar with the differences between these four techniques and the appropriate circumstances for their use. The exam will test your knowledge of how CPM networks are constructed, how schedules are computed, what the critical path is, and how networks are used to analyze and solve project scheduling, resource allocation, and resource leveling issues. The exam may also contain some rather elementary scheduling exercises. Variance analyses and earned value calculations can also appear in time management knowledge area questions.

EXAM TIP

Of the seven time management processes, six are in the planning process group; therefore, a large portion of the planning exam questions will come from this knowledge area. Be prepared to answer multiple questions dealing with estimating techniques, critical path, and network diagrams.

6

Things to Know

1. The seven processes of time management:
    *   **Plan Schedule Management**
    *   **Define Activities**
    *   **Sequence Activities**
    *   **Estimate Activity Resources**
    *   **Estimate Activity Durations**
    *   **Develop Schedule**
    *   **Control Schedule**
2. The purpose of the **schedule management plan**
3. **Analytical techniques** for the Plan Schedule Management process
4. **Activity attributes** and their relationship to the **WBS dictionary**

5. The four types of **dependencies**
   - **Mandatory**
   - **Discretionary**
   - **External**
   - **Internal**
6. **Network diagramming**, particularly the **precedence diagramming method**, and the four **logical relationships** (finish-to-start, start-to start, finish-to-finish, and start-to-finish)
7. **Lags** and **leads**
8. **Estimating tools and techniques** for activity duration:
   - **Expert judgment**
   - **Analogous estimating**
   - **Parametric estimating**
   - **Three-point estimating**
   - **Reserve analysis** and the **critical chain method**
9. The **critical path method** and **critical path method schedule development**
10. How to perform a **forward pass** and a **backward pass**
11. Definitions of **float**, **total float**, and **free float**
12. How to deal with **resource constraints** and **resource optimization techniques**, including **resource leveling** and **resource smoothing**
13. The **critical chain method**
14. The schedule compression techniques of **crashing** and **fast tracking** and how to apply them
15. The impact of **resource constraints** on a project schedule
16. How to **finalize resource requirements**
17. **Earned value terms and formulas** (these will be needed in several other sections too):
    - **Planned value** (PV)
    - **Actual cost** (AC)
    - **Earned value** (EV)
    - **Cost Variance** (CV)
    - **Schedule Variance** (SV)
    - **Cost performance index** (CPI)
    - **Schedule performance index** (SPI)
18. Key **interpersonal skills** for success

Key Definitions

Activity attributes: similar to a WBS dictionary because they describe the detailed attributes of each activity. Examples of these attributes are description, predecessor and successor activities, and the person responsible for completing an activity.

Contingency reserve: a dollar or time value that is added to the project schedule or budget and that reflects and accounts for risk that is anticipated for the project.

Crashing: using alternative strategies for completing project activities (such as using outside resources) for the least additional cost. Crashing should be performed on tasks on the critical path. Crashing the critical path may result in additional or new critical paths.

Critical path: the path with the longest duration within the project. It is sometimes defined as the path with the least float (usually zero float). The delay of a task on the critical path will delay the completion of the project.

Fast tracking: overlapping or performing in parallel project activities that would normally be done sequentially. Fast tracking may increase rework and project risk.

Float: the amount of time that a schedule activity can be delayed without delaying the end of the project. It is also called slack or total float. Float is calculated using a forward pass (to determine the early start and early finish dates of activities) and a backward pass (to determine the late start and late finish dates of activities). Float is calculated as the difference between the late finish date and the early finish date. The difference between the late start date and the early start date always produces the same value for float as the preceding computation.

Hammock: summary activities used in a high-level project network diagram.

6

Lag: the amount of time a successor's start or finish is delayed from the predecessor's start or finish. In a finish-to-start example, activity A (the predecessor) must finish before activity B (the successor) can start. If a lag of three days is also defined, it means that B will be scheduled to start three days after A is scheduled to finish.

Lead (negative lag): the amount of time a successor's start or finish can occur before the predecessor's start or finish. In a finish-to-start example, activity A (the predecessor) must finish before activity B (the successor) can start. A lead of three days means that B can be scheduled to start three days before A is scheduled to finish.

Logical relationships: there are four logical relationships between a predecessor and a successor:
- **Finish-to-start**, in which the predecessor must finish before the successor can start; this is the default relationship for most software packages
- **Finish-to-finish**, in which the predecessor must finish before the successor can finish
- **Start-to-start**, in which successor can start as soon as the predecessor starts
- **Start-to-finish**, in which the predecessor must start before the successor can finish; this is the least used and some software packages do not even allow it

Predecessor: the activity that must happen first when defining dependencies between activities in a network.

Project network schedule calculations: there are three types of project network schedule calculations—a forward pass, a backward pass, and float. A forward pass yields early start and early finish dates, a backward pass yields late start and late finish dates, and these values are used to calculate total float.

> **EXAM TIP**
> The most common logical relationship is the finish-to-start relationship. Use other relationships when there is a need for schedule compression.

Resource optimization techniques: techniques that are used to adjust the start and finish dates of activities that adjust planned resource use to be equal to or less than resource availability.

Schedule activity: an element of work performed during the course of a project. It is a smaller unit of work than a work package and the result of decomposition in the Define Activities process of project time management. Schedule activities can be further subdivided into tasks.

Scheduling charts: there are four types of scheduling charts—the Gantt chart, the milestone chart, the network diagram, and the time-scaled network diagram.
- A **Gantt chart** is a bar chart that shows activities against time; although the traditional early charts did not show task dependencies and relationships, modern charts often show dependencies and precedence relationships; these popular charts are useful for understanding project schedules and for determining the critical path, time requirements, resource assessments, and projected completion dates
- A **milestone chart** is a bar chart that only shows the start or finish of major events or key external interfaces (e.g., a phase kickoff meeting or a deliverable); a milestone consumes NO resources and has NO duration; these charts are effective for presentations and can be incorporated into a summary Gantt chart
- A **network diagram** is a schematic display of project activities showing task relationships and dependencies; the PDM is useful for forcing the total integration of the project schedule, for simulations and "what if" exercises, for highlighting critical activities and the critical path, and for determining the projected completion date
- A **time-scaled network diagram** is a combination of a network diagram and a bar chart; it shows project logic, activity durations, and schedule information

6

Standard deviation: the measurement of the variability of the quantity measured, such as time or costs, from the average. It is important to memorize these percentages.
- $\pm 1\sigma = 68.27\%$
- $\pm 2\sigma = 95.45\%$
- $\pm 3\sigma = 99.73\%$

This means that when you have $\pm 3\sigma$, 99.73% of all possible values fall within this range.

Statistical terms: the primary statistical terms are the project mean, variance, and standard deviation.
- The project **mean** (μ) is the sum of the means of the individual tasks
- The project **variance** is the sum of the variances of the individual tasks
- The project **standard deviation** (σ) is the square root of the project variance

Successor: the activity that happens second or subsequent to a previous activity when defining dependencies between activities in a network.

Triangular distribution or **three-point estimating**: takes the average of three estimated durations—the optimistic value, the most likely value, and the pessimistic value. By using the average of three values rather than a single estimate, a more accurate duration estimate for the activity is obtained.

Weighted three-point estimating or **beta/PERT**: the program evaluation and review technique (PERT) uses the three estimated durations of three-point estimating but weighs the most likely estimate by a factor of four. This weighted average places more emphasis on the most likely outcome in calculating the duration of an activity. Therefore, it produces a curve that is skewed to one side when possible durations are plotted against their probability of occurrence.

What-if scenario analysis: the process of evaluating scenarios in order to predict their effect on project objectives.

PLAN SCHEDULE MANAGEMENT PROCESS

The Plan Schedule Management process is the process of establishing the policies, procedures, and documentation for planning, developing, managing, executing, and controlling the project schedule. Planning for schedule management is important as an input to the overall **project management plan**.

The Plan Schedule Management process leverages information defined in the **project charter** and the **scope management plan**, such as the method for defining formal acceptance, and applies **expert judgement**, **analytical techniques**, and **meetings** to produce a singular output, the **schedule management plan**.

Schedule Management Plan

The schedule management plan establishes the criteria and activities for developing, monitoring, and controlling the schedule. Organizations may use very different methods and tools for developing a schedule for a project. The schedule management plan will typically include discussion and agreement on the following:
- Schedule development methods and tools
- Level of accuracy for activity durations
- Maintenance of project schedules
- Control thresholds that must be met before action needs to be taken

Analytical Techniques

There are many different methodologies, tools, and approaches that project teams can employ in establishing and maintaining schedules. The project team must evaluate these options and determine the best approach for its project and organization.

Rolling wave planning, leads and lags, and alternatives analysis are some of the many techniques that can be used during the Plan Schedule Management process.

> **EXAM TIP**
> Know the definitions of rolling wave planning, leads, lags, and alternatives analysis because they are all analytical techniques that can be used in planning.

6

DEFINE ACTIVITIES PROCESS

This process involves the identification of the specific activities that must be performed in order to produce the deliverables of the WBS **work packages**. Therefore, the key inputs are the **schedule management plan**, the **scope baseline**, including the **WBS** and **WBS dictionary**, the **enterprise environmental factors**, and the **organizational process assets**. Each of the WBS work packages is then broken down (decomposed) into smaller components called activities. These activities comprise the **activity list**. Note that **decomposition** is the tool and technique used in the Create WBS process as well.

The primary difference between decomposition in the Define Activities process and in the Create WBS process is the final output. In the Create WBS process, the results are deliverables; in the Define Activities process, the results are activities. In some areas, the WBS and activity list are developed concurrently, but in every case, by further breaking down the work, the scope is further refined and could result in WBS updates.

Activity Attributes and their Relationship to the WBS Dictionary

Activity attributes describe the characteristics of each activity planned within the work package. Similar to the WBS dictionary, the activity attributes typically include descriptions, predecessor and successor tasks, the person or department responsible for the task, and any other assumptions that may have been made in detailing what it takes to complete the activity. Figure 6-1 on the following page shows how activity attributes relate to the WBS dictionary.

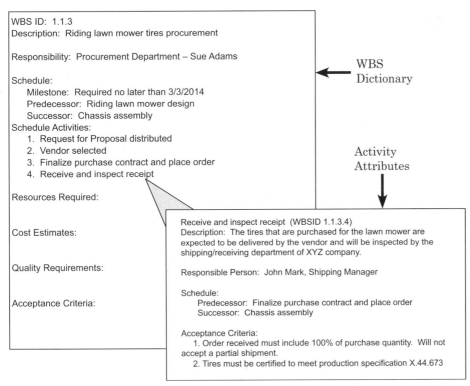

WBS ID: 1.1.3
Description: Riding lawn mower tires procurement

Responsibility: Procurement Department – Sue Adams

Schedule:
    Milestone: Required no later than 3/3/2014
    Predecessor: Riding lawn mower design
    Successor: Chassis assembly
Schedule Activities:
    1. Request for Proposal distributed
    2. Vendor selected
    3. Finalize purchase contract and place order
    4. Receive and inspect receipt

Resources Required:

Cost Estimates:

Quality Requirements:

Acceptance Criteria:

WBS Dictionary

Activity Attributes

Receive and inspect receipt (WBSID 1.1.3.4)
Description: The tires that are purchased for the lawn mower are expected to be delivered by the vendor and will be inspected by the shipping/receiving department of XYZ company.

Responsible Person: John Mark, Shipping Manager

Schedule:
    Predecessor: Finalize purchase contract and place order
    Successor: Chassis assembly

Acceptance Criteria:
    1. Order received must include 100% of purchase quantity. Will not accept a partial shipment.
    2. Tires must be certified to meet production specification X.44.673

Figure 6-1
WBS Dictionary and
Activity Attributes

SEQUENCE ACTIVITIES PROCESS

The four processes of Sequence Activities, Estimate Activity Resources, Estimate Activity Durations, and Develop Schedule are so closely linked that some projects combine the four processes into a single effort to come up with the project schedule. The Sequence Activities process involves organizing project tasks in the order they must be performed. For example, one cannot start the lawn mower mentioned in Figure 6-1 above until the ignition component is operational, and one cannot attach wheels until they are received from the supplier. This organization is accomplished by using the activity list and product description to determine the immediate predecessor(s) and successor(s) of each task. In addition, **mandatory dependencies** (hard logic), **discretionary dependencies** (preferred or soft logic), **external and internal dependencies**, and **milestones** must be considered.

EXAM TIP

Know how to calculate duration and elapsed time for simple PDM networks. PDM can also show lead (i.e. negative lag) and lag times.

Network Diagramming

Project schedule **network diagrams** are an important output of the Sequence Activities process. The main technique used is the **precedence diagramming method** (PDM). PDM is used for critical path determination. The network diagram may show that activity refinements are necessary, and it may result in updates to the activity list. There are many exam questions on critical path as well as on float, duration estimates, and forward and backward passes.

- In the precedence diagramming method, also called **activity-on-node** (AON), boxes or circles are used to represent activities while arrows show the sequence of workflow
- There are four types of dependencies or logical relationships—**finish-to-start**, **finish-to-finish**, **start-to-start**, and **start-to-finish**

Lags and Leads

Lags and leads are used for different purposes. A lag is used when there is a waiting period between a predecessor and a successor activity. Pouring concrete is a good example. When the activity of pouring concrete is complete, there is a waiting period before any of the next activities can be performed on that concrete. Assuming a three-day lag, the schedule relationship would be depicted as in Figure 6-2 below.

Figure 6-2
Lag

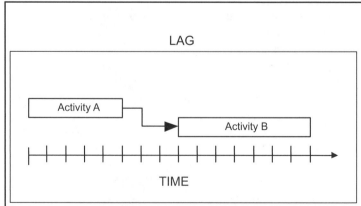

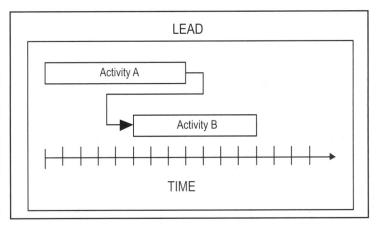

Figure 6-3
Lead

Leads are typically used when some acceleration is needed on the project schedule. In a software development project, for instance, ideally all training materials should be started after the completion of software testing. To accelerate the schedule by three days, the training material task could start three days earlier without testing being fully complete. This schedule relationship would be depicted as in Figure 6-3 above.

ESTIMATE ACTIVITY RESOURCES PROCESS

Once the project schedule **network diagram** is available, resources are assigned. Resources come in several forms—people, equipment, supplies, materials, facilities, and money. One of the key components of determining an accurate schedule for any project is the ability to determine accurately the type of resources (by description, quantity, and availability) needed to perform work activities. These resource estimates can then be used to estimate project costs. Therefore, the Estimate Activity Resources process is very closely tied to the Estimate Costs process in the next chapter.

The *PMBOK® Guide* breaks down estimating into two steps. The first is estimating resources from a skill and availability standpoint (which occurs in the Estimate Activity Resources process) and the second is estimating the duration of each work package based on the decisions made in the Estimate Activity Resources process.

6

In addition to the **schedule management plan**, the **activity list**, **activity attributes**, **activity cost estimates**, **resource calendars**, and **risk register** are all key inputs to the Estimate Activity Resources process. By assigning resources and estimates to each activity (activities which then are rolled up to work packages within the WBS), a more detailed and accurate schedule is produced. The primary outputs of this process are the activity resource requirements, the updates to project documents, and the **resource breakdown structure**. The resource breakdown structure is a hierarchical structure of the identified resources broken out by category and type. It is a useful document for managing resource utilization information.

ESTIMATE ACTIVITY DURATIONS PROCESS

Once the project tasks have been organized in the order that they are to be performed and the resources have been identified, the Estimate Activity Durations process can take place. This is the process of defining the number of work periods that will be needed to complete the individual activities. In estimating the duration of a task, it is necessary to know the resource requirements and capabilities, factors influencing elapsed time, and any historical information of similar tasks. The project team must decide if the types of estimates should be **deterministic** (if a single estimate is used and the duration is known with a fair degree of certainty) or **probabilistic** (if the task duration is uncertain, a three-point estimating technique or weighted average is used).

The primary tools and techniques in this process are the various estimating methods. In obtaining the duration estimates of the tasks, activity refinements may be necessary and updates may need to be made to the activity attributes.

Estimating Tools and Techniques

Expert judgment involves using one or more subject matter experts as a source for obtaining estimates. If subject matter experts lack historical information or experience, the risk of inaccurate estimates is increased.

Within time management, the project manager may consult with a senior resource (an "expert") in order to obtain good work effort estimates rather than have inexperienced team members provide them.

Analogous estimating (or **top-down estimating**): obtained by comparing the current project activities to previous project activities and using the actual duration of the previous similar activity as the basis for the current activity. The degree of similarity affects accuracy. This technique should be used early in the estimating cycle when there is not much detail known about the activity. Analogous estimating is considered to be a form of expert judgment.

Parametric estimating (sometimes called **quantitatively-based estimating**): the quantities of the units of work are multiplied by a productivity unit rate to obtain an estimated activity duration. These estimates are based on historical information that has been codified. An example of this type of estimating would be the creation of a training manual in terms of pages per day. A junior writer might produce two pages per day while a senior writer might produce five pages per day. A software development example would be the number of lines of code produced per day by various levels of programmers.

Three-point estimating: a technique to reflect risk in the estimates that are provided for both time and cost. The two formulas that are typically used to adjust for risk are triangular and Beta/PERT weighted average estimates.

- With A = lowest value, B = highest value, M = most likely value, and V = variance for a task
  $V = \sigma^2$ or
  $V = [(A - B)^2 + (M - A)(M - B)] \div 18$
- Triangular Mean (μ)
  $\mu = (A + M + B) \div 3$
- PERT or weighted average mean
  $\mu = (A + 4ML + B) \div 6$
- Standard deviation (σ)
  $\sigma = (B - A) \div 6$

Reserve analysis and the **critical chain method**: depending on the nature of the project duration estimates, reserve analysis may include additional time reserves or buffers. A project with high schedule risk would contain a bigger contingency reserve than a project with little schedule risk. A key aspect of the critical chain method uses feeding buffers and an overall project buffer to protect target dates along the critical chain from slippage. These **contingency reserves** or **buffers** should be analyzed regularly and adjusted as more project data become available.

To increase the probability of schedule acceptance, it is important for the project manager to involve the people who are responsible for the activity delivery in creating the duration estimates. Engaging team members and utilizing **group decision-making techniques** will increase estimate accuracy and team commitment.

DEVELOP SCHEDULE PROCESS

The Develop Schedule process utilizes the duration estimates for each task in order to determine the start and finish dates for each task. By traversing the longest path within the project, the **critical path** can be determined and the project schedule can be developed. Therefore, necessary inputs to the Develop Schedule process are the **network diagrams** with **duration estimates**, **resource requirements**, the **activity list**, **resource calendars**, the **project scope statement**, **enterprise environmental factors**, and **organizational process assets**.

In order to produce the project schedule, schedule network analysis is used to calculate early and late start and finish dates without regard for resource availability and limitations. Schedule network analysis uses various analytical techniques such as the **CPM**, the **critical chain method**, **resource optimization**, **modeling techniques**, **schedule compression**, and scheduling tools. Loops and open-ended nodes in the network diagram are adjusted before the analytical technique is applied.

Float, or **slack**, is a concept you must understand. You must know how it presents challenges and opportunities to project schedulers.

The **project schedule**, **schedule baseline**, and updated schedule management plan are the primary outputs of the Develop Schedule process.

Critical Path Method Schedule Development

When calculating the critical path for a project, you start with the network diagram and duration estimates to determine:
- Float
- Early start date (ES)
- Early finish date (EF)
- Late start date (LS)
- Late finish date (LF)

Early Start	Duration	Early Finish
	Task Name	
Late Start	Slack	Late Finish

Figure 6-4
Matrix for a
Network Diagram

These variables are formatted into a simple matrix for each task of the network diagram, as shown in Figure 6-4.

The network diagram is traversed from left to right in a **forward pass** to determine ES and EF:
- Start time of the project is zero (0)
- ES = EF of the immediate predecessor
- EF = ES plus duration
- A successor starts when ALL its predecessors are complete
- Project finish date = EF of the final task

Figure 6-5 illustrates the results of a forward pass.

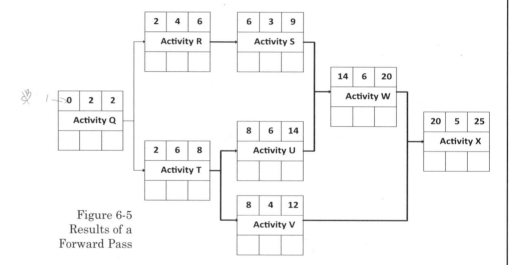

Figure 6-5
Results of a
Forward Pass

Once the forward pass is complete, begin the **backward pass** by traversing the network diagram from right to left to determine LF and LS of each task:
- Project LF = EF of the final task of the project
- LS = (LF minus duration) + 1
- LF = the earliest LS of ALL its successors

Once LF and EF are known, **float** can be calculated:
- Float = LF – EF

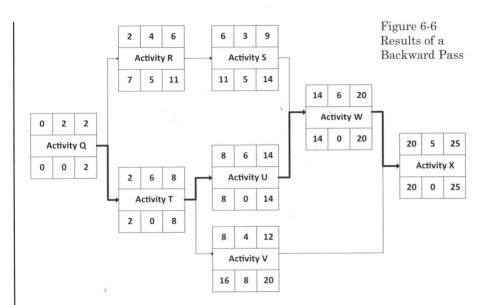

Figure 6-6
Results of a
Backward Pass

The results of a backward pass are shown in Figure 6-6. Note that both activities R and S in the example each have a **total float** of five days. The combined duration of activity R and S is seven days, while the combined duration of activity T and U (on the critical path) is twelve days, and activity W (also on the critical path) is the successor to both activities U and S. Free float is defined as the amount of time an activity can be delayed without delaying the start of an immediate successor. Look at activities R and S. Both of them show a float of 5 days, but each do not have 5 days of free float. Only if activity R is NOT delayed does activity S have 5 days of free float.

In CPM analysis, a finish-to-start relationship is used exclusively. Once the network diagram is completed and the schedule activity relationships are known, a **Gantt chart** can be developed. Gantt charts are generally used to monitor schedule progress. Other relationships besides finish-to-start can be reflected along with leads and lags. A sample Gantt chart is shown in Figure 6-7 on the following page.

EXAM TIP

Resource requirement updates could lengthen a project schedule. If an activity requires three people working for four calendar weeks but only two people are available, it will take six calendar weeks to complete the project, all other factors being equal.

ID	Task Name	Start	Finish	Duration	Jul 2005							
1	Activity Q	6/13/2005	6/14/2005	2d								
2	Activity R	6/15/2005	6/20/2005	4d								
3	Activity S	6/21/2005	6/23/2005	3d								
4	Activity T	6/15/2005	6/22/2005	6d								
5	Activity U	6/23/2005	6/30/2005	6d								
6	Activity V	6/23/2005	6/28/2005	4d								
7	Activity W	6/30/2005	7/7/2005	6d								
8	Activity X	7/8/2005	7/14/2005	5d								

Figure 6-7
Gantt Chart

Case Study Exercises

Exercise 6-1: Using the WBS for the Lawrence Garage Project developed in Chapter 5, Exercise 5-2, create a network diagram showing the dependencies of the various WBS items. Use the activity-on-node technique. Also include your estimates for how many days each activity will take. All estimates should be in whole days.

Exercise 6-2: Draw a Gantt chart based on the network diagram from Exercise 6-1 for the Lawrence Garage Project.

Dealing with Resource Constraints and Resource Optimization Techniques

In many cases projects are resource constrained, which limits optimal scheduling. Three methods that can be used to facilitate scheduling with limited resources are resource leveling, resource smoothing, and the critical chain method.

Resource Leveling

Resource leveling is a schedule network analysis technique that is performed after the critical path has been determined to address specific delivery dates and to take into account resource availability or to keep resource usage at a constant level during specified time periods of the project. The resulting schedule often has an altered critical path and could result in the project taking longer to complete.

The resource leveling technique reviews the project schedule for either over- or under-allocation of resources. There may be times when more resources are needed than are available, which could occur for equipment, people, or subject matter experts. There may be individual resources that are assigned to one or more tasks during a time period that exceed an individual's available time.

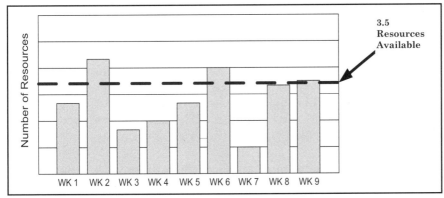

Figure 6-8
Resource
Histogram
with Leveling

Resource usage is evaluated at the project level by preparing a **resource histogram** to determine total resources used in each time period. The project manager may try to eliminate spikes and gaps in total resource usage within the histogram, as shown in Figure 6-8. Resource leveling is used to reschedule concurrent activities in which individuals have one or more constraints. Resource leveling may result in extending the duration of the project.

Figure 6-8 on the previous page shows a histogram of a project in which there are 3.5 resources available. You can see that there are two work periods in which there are not enough resources for the work to be performed, while in other periods there are ample resources.

Resource Smoothing

Resource smoothing differs from **resource leveling** in that it attempts to adjust activities within the existing free and total float, thereby minimizing the potential for affecting the **critical path**.

Critical Chain Method

The critical chain method is a schedule network analysis technique that is performed after the critical path has been determined to take into account resource availability. If a resource is unavailable for any period of time, the resulting schedule often has an altered critical path.

The critical chain method is an alternative to the critical path method. It views the project as a system instead of a network of independent tasks. While the **critical path** is determined by task dependency relationships, the critical chain adds resource dependencies to define a resource-limited schedule. The longest sequence of resource-leveled tasks is the critical chain.

The critical chain method tries to adjust for problems in estimating and managing tasks that result from:
- Poor multi-tasking
- Estimates with too much contingency for uncertainty
- Work that expands to fill the available time
- Waiting until the latest possible time to start
- A lack of prioritization

EXAM TIP

Remember that the longest sequence of resource-leveled tasks is the critical chain.

The critical chain method includes techniques that:
- Focus on the project end date, not individual task completion
- Use an estimate for tasks at the 50% confidence level, so that half the tasks will be earlier than the estimate and half the tasks will be later
- Level the tasks throughout the whole project to determine the critical chain
- Eliminate multi-tasking on significant tasks while focusing on critical chain tasks
- Create project, feeding, and resource buffers for project uncertainties instead of building contingency into individual tasks; uncertainty is determined by comparing the difference between estimates at a high confidence level (90% to 95%) and the 50% confidence level
- Establish feeding buffer time at the end of a series of tasks that feed into the critical chain
- Establish project buffer time at the end of the critical chain to protect the project end date
- Monitor the use of buffer durations rather than individual task performance

Schedule Compression

Schedule compression is a special case of schedule network analysis in which the project schedule is shortened without changing the project scope. **Crashing** and **fast tracking** are two techniques to accelerate the project schedule without changing the project scope. Both techniques require an understanding of network diagrams.

Both crashing and fast tracking can result in increased costs and can also increase the risks of not achieving the shorter project schedule. Analysis of a **Monte Carlo simulation** and **what-if scenario analysis** are modeling techniques that can be used to calculate the probable results for the total project.

Possible Opportunity	Schedule	Cost	Risk
Reduce Activity W 3 days by hiring an expert	Reduce 3 days		Low
Break Activity W into 2 equal activities and hire another resource at the current rate	Reduce 3 days		High
Reduce Activity T 2 days by working over the weekend	Reduce 2 days		Medium

Figure 6-9
Schedule
Compression

As a project manager you must recognize that there are always tradeoffs to schedule compression. You may have to bring in extra resources or more expensive resources in order to perform schedule activities in shorter time frames or in parallel.

Depending upon **risk tolerances** and the **constraints** and **assumptions** in your scope statement, the choice of where to compress the schedule will usually become obvious.

Figure 6-9 shows an example of opportunities to compress the project schedule depicted in Figures 6-5 through 6-7.

Finalizing Resource Requirements

The output of the Develop Schedule process is an official set of plans for the project that includes:
- The project schedule
- The schedule baseline
- Schedule data which may include a resource requirement histogram, such as in Figure 6-8
- Project document updates
- Project calendars
- Project management plan updates

The steps to be taken for assigning people to work each schedule activity in a work package and detailed into a Gantt chart are:
- Assign resources to each activity, one resource at a time
- For each time period, sum up the resources required
- Develop a resource histogram

CONTROL SCHEDULE PROCESS

The Control Schedule process involves the handling of factors that could impact the project schedule. It involves:
- The review and approval processes of changes that lead to schedule changes
- Determining the size and impact of each change
- Managing actual schedule changes

Changes to the project schedule could come in the form of a **change request** or in response to project performance or nonperformance needing corrective action. **Performance reviews**, **resource optimization**, and **modeling techniques** are key to schedule control. Comparing actual start and finish dates to planned dates allows you to find problems early on in a project.

Although **earned value management** (EVM) is not discussed in detail in project time management, it is one of several useful techniques in **performance reviews**, and many exam questions utilize earned value concepts in this knowledge area.

Earned Value Terms and Formulas

EVM is a technique that describes plans and performance in terms of monetary amounts to calculate both the cost and schedule status of a project that has started. Since costs are the basis of all the calculations, even schedule information is stated in cost units. Also, remember that EVM is always as of a specific date (the data date).

Three dimensions of EVM are planned value, actual cost, and earned value.

- **Planned value** (PV) is the sum of the approved cost estimates for activities scheduled to be performed during a given period
- **Actual cost** (AC) is the amount of money actually spent in completing work in a given period
- **Earned value** (EV) is the sum of the approved cost estimates for activities completed during a given period

Variables monitored in EVM are cost and schedule variances, the cost and schedule performance indices, and budget and estimate at completion.

- **Cost variance** (CV) is earned value (EV) minus actual cost (AC); it is the difference between the budgeted cost of the work completed and the actual cost of completing the work; a negative number means the project is over budget
  - $CV = EV - AC$
- **Schedule variance** (SV) is earned value (EV) minus planned value (PV); it represents the difference between what was accomplished and what was scheduled; a negative number means the project is behind schedule
  - $SV = EV - PV$
- **Cost performance index** (CPI) is earned value (EV) divided by actual cost (AC); it is the ratio of what was completed to what it cost to complete it; values less than 1.0 indicate we are getting less than a dollar's worth of value for each dollar we have actually spent; CPI measures cost efficiency
  - $CPI = EV \div AC$
- **Schedule performance index** (SPI) is earned value (EV) divided by planned value (PV); it is the ratio of what was actually completed to what was scheduled to be completed in a given period; values less than 1.0 mean the project is receiving less than a dollar's worth of work for each dollar that was scheduled to be spent; SPI measures schedule efficiency
  - $SPI = EV \div PV$

EXAM TIP

EV is always the first variable to the right of the equal sign in EVM. For example:

$CV = EV - AC$
$SV = EV - PV$
$CPI = EV \div AC$
$SPI = EV \div PV$

KEY INTERPERSONAL SKILLS FOR SUCCESS

The interpersonal skill highlighted in this chapter is:

Conflict Management

Because of the challenges of managing projects, conflict is expected at some time during the project life cycle. Conflict may arise from differences over schedules, technical requirements, cost, personalities, organizational challenges, management, the customer or end user, and a wide variety of other causes.

Project managers who successfully lead teams find ways to encourage conflict when it can lead to better project decisions and to resolve conflict when necessary to create a collaborative approach among team members. Managing conflict requires that the project manager finds a way to build and maintain a trusting relationship among team members and others associated with the project.

Much of the effort in developing a detailed and comprehensive schedule will inevitably cause conflict within the team and especially with the customer. Initial expectations of project duration are generally too optimistic. Detailing out a schedule usually requires that these expectations be modified.

SAMPLE PMP EXAM QUESTIONS ON TIME MANAGEMENT

1. You are trying to estimate the duration of a project activity related to your annual member conference. One of your team members has completed the same activity for the prior 5 conferences. She estimates that it will take about 60 hours, but that it could be as little as 50 hours or as much as 100 hours. What estimate is most reasonable to use in your project plan?

   a) 60
   b) 65
   c) 70
   d) 100

2. The main output of the Estimate Activity Resources process is:

   a) Analogous estimate
   b) Activity resource requirements
   c) Bottom-up estimate
   d) Expert judgment

Notes:

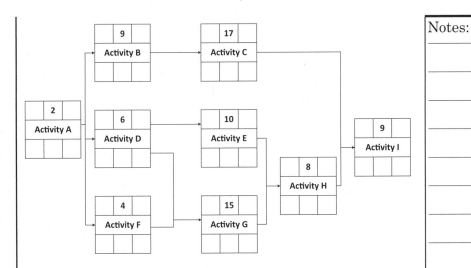

The durations above are in weeks.

3. What is the critical path of the project in the chart above?

   a) A-B-C-I
   b) A-D-E-H-I
   c) A-D-G-H-I
   d) A-F-G-H-I

4. Your project is represented above in question 3. Jason, a team member who is working on Activity C, has won a trip to Machu Picchu. You want the activity to be completed before he leaves on his trip. What is the earliest he can leave?

   a) At the beginning of week 11
   b) At the beginning of week 14
   c) At the end of week 31
   d) At the end of week 28

Notes:

5. You are the project manager for a business process improvement project for a strategic business process. A key resource on the project, William, has been asked by the CEO to work on another project for the next two weeks. William is scheduled to start work on key deliverables in two weeks. Your response to this request should be:

a) Communicate to the CEO that William is not available because he is critical to the success of the project
b) Ask William to turn down the CEO's project because it may take too long
c) Identify the issue as a risk and add two weeks to the project as a contingency plan
d) Assume the CEO's project will take only two weeks and keep the schedule the same

6. You want to control the schedule closely for your project to exhibit at a trade show. It can't be late! To make sure you are tracking progress, you will establish rules of performance measurement such as:

a) Level of detail for the work breakdown structure
b) Cost variance and cost performance index thresholds
c) Control accounts where you will take actual cost measurements
d) Earned value measurement techniques

7. As a control tool, the bar chart (Gantt) method is most beneficial for:

a) Rearranging conflicting activities
b) Depicting actual versus planned activities
c) Showing the outer dependencies of activities
d) Identifying when activities are in control

Notes:

8. How does effort differ from duration?

   a) Effort represents the labor units required, while duration represents the work units required
   b) Effort is used to create the project schedule, while duration is used to create the project budget
   c) Effort is mainly affected by the number of resources available, while duration is not
   d) Effort is mainly affected by the availability of resources, while duration is affected by the difficulty of the work

9. When an activity must be completed before another activity can start, the logical relationship is called:

   a) Start-to-finish
   b) Finish-to-start
   c) Start-to-start
   d) Finish-to-finish

10. As the project manager, you are scheduling some activities in which there is a delay between the finish of the predecessor activity and the start of the successor activity. These dependencies would be reflected in the schedule with a:

   a) Finish-to-start relationship
   b) Lag
   c) Start-to-finish relationship
   d) Lead

Notes:

6

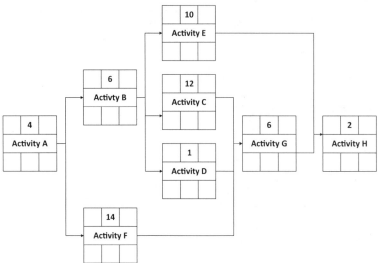

The durations above are in days.

11. In the chart above, if you crash the schedule and reduce Activity C by 3 days, what is the effect on the project?

    a) A new critical path is created: A-B-E-G-H
    b) The duration of the project will be extended
    c) The critical path is not affected
    d) Activity C can't be crashed because it's not on the critical path

12. The agreed amount of variation to be allowed before some action needs to be taken is described in the schedule management plan as:

    a) Units of measure
    b) Level of accuracy
    c) Control thresholds
    d) Rules of performance measurement

Notes:

13. You are trying to estimate activity resources, and you want to find out which resources have the required skill and available time. What document should you look at first?

    a) Activity list
    b) Resource calendars
    c) Risk register
    d) Activity cost estimates

14. One form of progressive elaboration used in the Define Activities process is:

    a) Monte Carlo simulation
    b) The organizational breakdown structure
    c) Rolling wave planning
    d) WBS dictionary

15. You and your team have completed the WBS and want to decompose the work into more detail for purposes of estimating time and cost, developing a sequence of work, identifying risk, and performing a make vs. buy analysis. Where would you document these details?

    a) Schedule management plan
    b) Work breakdown structure
    c) Rolling wave plan
    d) Activity list

Notes:

6

ANSWERS AND REFERENCES FOR SAMPLE PMP EXAM QUESTIONS ON TIME MANAGEMENT

Section numbers refer to the *PMBOK® Guide*.

1. **B   Chapter 6.5.2.4 – Planning**
    A) This is the most likely value, but it doesn't take into consideration the best case and worst case estimates; B) this value is computed by using the Beta/PERT formula = (optimistic + 4 (most likely) + pessimistic) ÷ 6; this takes into consideration the team members' past experience; C) this is a simple average of the 3 data points; D) no organization can afford to estimate using the worst case scenario.

2. **B   Section 6.4.3.1 – Planning**
    A) and C) are outputs of the Estimate Activity Durations process; D) is a technique used in many time management processes.

3. **C   Section 6.6.2.2 – Planning**
    The critical path is the longest path through the network; A) is 37; B) is 35; C) is 40; D) is 38.

4. **D   Section 6.6.2.2 – Planning**
    The early finish date is week 28.

5. **D   Section 6.7.2.1 – Monitoring and Controlling**
    A) may be a career-limiting move; B) may be a career-limiting move for William and he may be resentful; C) identification as a risk is a good thing, but at this time there may not be any need to add schedule contingency even though you may want to look at some resource contingency; D) it appears that the schedule won't have to be impacted, but you'll want to keep an eye on the progress of the CEO's project.

6. **D   Section 6.1.3.1 – Planning**
    A) is required for defining activities, but much more is needed to control schedule; B) and C) relate to cost, not schedule.

7. **B** **Section 6.7.1.2 – Monitoring and Controlling**
Gantt (bar) charts show start and end times of activities. They are useful for portraying actual versus planned start and end times.

8. **A** **Section 6.5 – Planning**
Effort represents actual time worked; it is a basis for estimating duration. Duration may be shorter than effort if you can add resources, and it may be longer than duration if resources have limited availability or there is wait time.

9. **B** **Section 6.3.2.1 – Planning**
The most common relationship is the finish-to-start relationship.

10. **B** **Section 6.3.2.3 – Planning**
Although the activities described have a finish-to-start relationship, the delay in the start of the successor is a lag.

11. **C** **Section 6.6.2.2 – Planning**
If the duration of Activity C is shortened by 3 days, (i.e. from 12 days to 9 days), then the critical path will remain the same: A-B-D-G-H.

12. **C** **Section 6.1.3.1 – Planning**
These are all potential components of the schedule management plan.

13. **B** **Section 6.4.1.4 – Planning**
These are all inputs to the Estimate Activity Resources process.

14. **C** **Section 6.2.2.2 – Planning**
Rolling wave planning entails waiting until the deliverable or subproject is clarified before developing the details of the WBS.

15. **D** **Section 6.2.3.1 – Planning**
The activity list should include a unique identifier and enough detail so that the project team knows what work needs to be done.

CASE STUDY SUGGESTED SOLUTIONS

Exercise 6-1
Network Diagram for the Lawrence Garage Project
This example uses the phases of the project as tasks to
show dependencies.

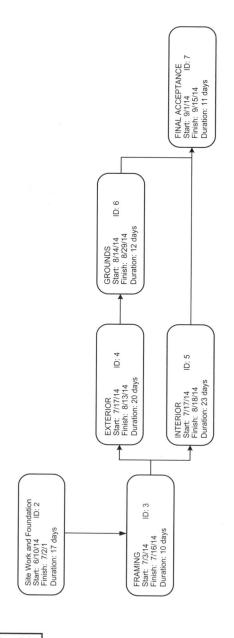

Exercise 6-2
Gantt Chart for the Lawrence Garage Project

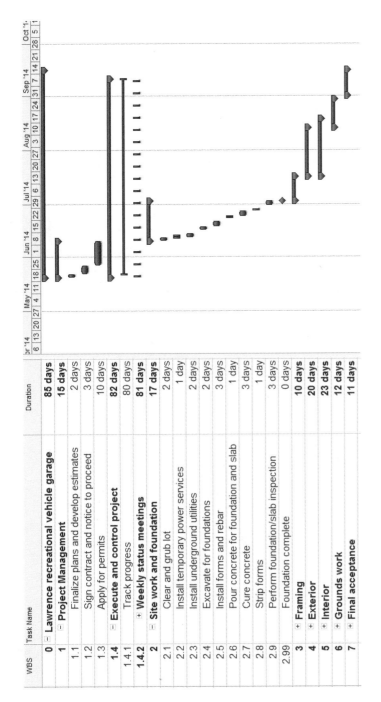

WBS	Task Name	Duration
0	Lawrence recreational vehicle garage	85 days
1	Project Management	15 days
1.1	Finalize plans and develop estimates	2 days
1.2	Sign contract and notice to proceed	3 days
1.3	Apply for permits	10 days
1.4	Execute and control project	82 days
1.4.1	Track progress	80 days
1.4.2	Weekly status meetings	81 days
2	Site work and foundation	17 days
2.1	Clear and grub lot	2 days
2.2	Install temporary power services	1 day
2.3	Install underground utilities	2 days
2.4	Excavate for foundations	2 days
2.5	Install forms and rebar	3 days
2.6	Pour concrete for foundation and slab	1 day
2.7	Cure concrete	3 days
2.8	Strip forms	1 day
2.9	Perform foundation/slab inspection	3 days
2.99	Foundation complete	0 days
3	Framing	10 days
4	Exterior	20 days
5	Interior	23 days
6	Grounds work	12 days
7	Final acceptance	11 days

COST

CHAPTER 7 | **COST**

7

COST MANAGEMENT

Cost management questions on the PMP exam do not require you to be a math whiz. The questions address cost management from a project manager's perspective, which is much more general. However, these questions are NOT easy.

This section addresses a broad range of cost issues such as financial projections, cost estimating and budgeting, earned value, creating and interpreting S-curves, and forecasting. This is probably the section people spend the most time studying.

In addition, work performance information, earned value analysis, and communications are integrated into questions on cost. Formulas for figuring costs are fairly simple, but you must know them. You must also understand and be able to evaluate performance using earned value formulas and **variance analysis**.

Cost management is a three-step process comprised of estimating costs, determining a budget, and controlling costs.

Things to Know

1. The four processes of cost management:
   - **Plan Cost Management**
   - **Estimate Costs**
   - **Determine Budget**
   - **Control Costs**
2. **Cost management general concepts**
3. How to create a **cost management plan**
4. The difference between **estimating** and **pricing**
5. The **types of estimates**:
   - **Analogous estimating**
   - **Parametric estimating**
   - **Bottom-up estimating**
   - **Three-point estimating**

6. **Cost estimating techniques**:
   - **Reserve analysis**
   - **Vendor bid analysis**
   - **Ranges of estimates**
7. The importance of **group decision making**
8. What a **reserve analysis** is
9. **Funding requirements** and **funding limit reconciliation**
10. How to create a **cost baseline**
11. How to create **S-curves**
12. **Crashing** with time and/or cost tradeoffs
13. Earned value terms and formulas:
    - **Planned value** (PV)
    - **Actual cost** (AC)
    - **Earned value** (EV)
    - **Cost variance** (CV)
    - **Schedule variance** (SV)
    - **Cost performance index** (CPI)
    - **Schedule performance index** (SPI)
14. **Forecasting techniques**:
    - **Budget at completion** (BAC)
    - **Estimate at completion** (EAC)
    - **Estimate to complete** (ETC)
    - **Variance at completion** (VAC)
    - **To-complete performance index** (TCPI)
15. The **profitability measures**:
    - **Return on sales** (ROS)
    - **Return on investment** (ROI)
    - **Return on assets** (ROA)
    - **Present value** (PV)
    - **Net present value** (NPV)
    - **Internal rate of return** (IRR)
    - **Benefit cost ratio** (BCR)
    - **Payback method**

EXAM TIP

Crashing should be performed on activities on the critical path.

7

Key Definitions

Activity contingency reserve: budget for a specific WBS activity within the cost baseline that is allocated for identified risks that are accepted and for which contingent or mitigating responses are developed.

Contingency reserve: budget within the cost baseline or performance measurement baseline that is allocated for identified risks that are accepted and for which contingent or mitigating responses are developed.

Crashing costs: costs incurred as additional expenses above the normal estimates to speed up an activity.

Direct costs: costs incurred directly by a project.

Fixed costs: nonrecurring costs that do not change if the number of units is increased.

Indirect costs: costs that are part of doing business and are shared among all ongoing projects.

Management reserve: a dollar value, not included in the project budget, that is set aside for unplanned changes to project scope or time that are not currently anticipated.

Opportunity costs: costs of choosing one alternative over another and giving up the potential benefits of the other alternative.

Percent complete: the amount of work completed on an activity or WBS component.

Sunk costs: money already spent; there is no more control over these costs. Since these are expended costs they should not be included when determining alternative courses of action.

Variable costs: costs that increase directly with the size or number of units.

7

Cost Management General Concepts

The cost management of a project focuses on the cost of the resources needed to complete a project. The *PMBOK® Guide* emphasizes **life cycle costing** as a way to get a broader view of project costs. Life cycle costing includes acquisition, operation, maintenance, and disposal costs. Project decisions should take into consideration life cycle costing. For example, a project manager may purchase a proprietary technology because he or she realizes, in considering life cycle costing, that the purchase of the technology will decrease future maintenance support and training.

PLAN COST MANAGEMENT PROCESS

The Plan Cost Management process establishes the policies, procedures, and documentation for planning, managing, expending, and controlling costs. Planning for cost management is important as an input to the overall **project management plan**.

The key benefit of this process is that it directs how project costs will be managed throughout the project. Project costs are extremely hard to control, and having a structure around the approach to cost control is imperative.

Cost Management Plan

The cost management plan defines the processes that the project team will follow for:
- Units of measure
- Level of precision
- Level of accuracy
- Control thresholds
- Rules of performance measurement
- Linking control accounts to the company's financial reporting system

ESTIMATE COSTS PROCESS

There is often confusion between **cost estimating** and
pricing. In the Estimate Costs process, the costs of key
resources are estimated. The project team also considers
the tradeoffs between scope, time, and cost and evaluates
the overall impact of cost on the project. Pricing is a
business decision about what the customer or client
should be charged for a product or service produced by
the project. Cost is part of the pricing decision, but it is
only one component.

There are several types of estimates and cost estimating
techniques that the *PMBOK® Guide* outlines and are
described further in this chapter. The outputs of this
process are the **cost estimates** for each schedule activity,
the **basis for estimates**, and any updates to project
documents.

Types of Estimates

Analogous estimating is based on actual costs of
previous, similar projects and on comparing those
costs to current project work packages. The degree of
similarity between a prior project and the current
project affects the accuracy of the estimate. Also,
remember to consider changes in labor rates, purchased
resources, and overhead that have occurred since the
prior project was completed. Analogous estimating takes
less time and costs less than other types of estimates.

Parametric estimating is based on the statistical
relationship between historical information and other
variables. An example used in home construction is the
cost per square foot. This number remains about the
same no matter how big the house. Software
development companies may base their parametric
estimates on lines of code and function points.

Bottom-up estimating is based on a very detailed
work breakdown structure that allows the estimator to
estimate each work package more accurately. When the
individual work package estimates are rolled up, they

become the project estimates. This technique requires detailed specifications as well as a good understanding of the cost components for each work package.

Three-point estimating is used to take risk into account. One method is Beta/PERT, which estimates optimistic (O), pessimistic (P), and most likely (ML) values for costs. The Beta/PERT technique may be used to calculate expected activity cost using the formula:
- Expected Cost $= (O + 4ML + P) \div 6$

Cost Estimating Techniques

Reserve analysis is a key technique for estimating costs since estimates are highly sensitive to the expected monetary value of risk events. Risk assessment is used to determine the appropriate level of contingency reserves. Take note of the risk analysis discussed in Chapter 11.

Vendor bid analysis can occur when a project or items within a project are procured. The team may be required to estimate and analyze what a project or procured item should cost.

Ranges of estimates are on an order of magnitude. As more detailed planning information becomes available, cost estimates should be revised. These revisions may occur at the end of a major phase as well. When starting the project, a **rough order of magnitude** (ROM) estimate may have a range of ± 50%. Detailed planning may allow the project manager to reduce the range to ± 10%.

Group Decision Making

Group decision making techniques are valuable not only in estimating activity durations but also in estimating costs. The theory here is that the more you include individuals in the decision making process, the more likely you will be to see improved task estimates and a commitment to those estimates.

> **EXAM TIP**
> Project costs can only be estimated with 100% certainty at the conclusion of a project.

7

DETERMINE BUDGET PROCESS

Once the cost estimates for each activity are available, the Determine Budget process involves aggregating the values to create an authorized cost performance baseline for the cost management plan and to assess risks defined in the risk register. This cost baseline is used to measure project performance.

Reserve Analysis

Reserve analysis is conducted during the Determine Budget process as well as during the Estimate Costs process. Both contingency and management reserves may be determined for the project.

- **Contingency reserves** are allowances established to account for uncertainty and potential risk events
- **Management reserves** may be established for unplanned changes to cost and are not included as part of the cost baseline

Funding Requirements and Funding Limit Reconciliation

Similar to home construction or remodeling projects, funds for construction are released as certain milestones are achieved. Many organizations have limited resources and expenditures which must be managed. Based on funding limits that may be set by a group such as the customer or finance department, work schedules may have to be adjusted or reconciled to accommodate the payment schedule.

Cost Baseline

The cost baseline is the authorized **budget at completion** (BAC) spread over the project schedule. It is the plan used to compare to actual costs in order to assess project process. Cumulative costs are shown as an S-curve.

> **EXAM TIP**
> Study Figure 7-8 in the *PMBOK® Guide*; this figure visually demonstrates the various types of contingencies and how they relate to the project budget.

7

S-Curves

An S-curve is a graphical representation of accumulated budgeted costs over time. Typically, costs rise gradually early in the project, accelerate during execution, and taper off as a project closes, creating a curve that resembles an "S" as shown in Figure 7-1.

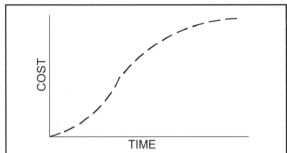

Figure 7-1
Sample S-Curve

An example of the data that could generate an S-curve could be the weekly labor costs to create project deliverables such as the WBS, project schedule, and histogram. Next, add in any material costs and factors for overhead items. Figure 7-1 is an example of an S-curve like one that would be generated from the data in Figure 7-2.

Figure 7-2
S-Curve data

Activity	Week 1	Week 2	Week 3	Week 4	Week 5
Q	400				
R	600	200			
S		600			
T	600	600			
U		200	800		
V		200	400		
W			200	1,000	
X					1,000
Total Labor	1,600	1,800	1,400	1,000	1,000
Material	1,000	1,000	1,000	1,000	1,500
Overhead (10% of labor)	160	180	140	100	100
Total Budget	2,760	2,980	2,540	2,100	2,600
Cumulative Budget	2,760	5,740	8,280	10,380	12,980

Case Study Exercise

Exercise 7-1: Using the Gantt chart from the previous exercises and the budget data below for the Lawrence Garage Project, develop an S-curve.

Work Package	Labor	Materials and Equipment	Total
Project Management	$5,000		$5,000
Site Prep	$500	$500	$1,000
Foundation	$1,000	$1,500	$2,500
Framing	$2,500	$7,500	$10,000
Utilities			
Plumbing (subcont.)			$800
Electrical (subcont.)			$1,100
HVAC (subcont.)			$1,800
Roof (subcont.)			$4,500
Stucco (subcont.)			$500
Insulation (subcont.)			$550
Drywall	$350	$1,950	$2,300
Painting	$250	$250	$500
Interior Finish	$150	$450	$600
Driveway and Walks	$300	$1,200	$1,500
Acceptance	$225		$225
TOTALS	$10,275	$13,350	$32,875

CONTROL COSTS PROCESS

Projects often operate under many constraints, but of all the constraints, cost is one of the most difficult to control. As changes occur, adjustments and contingency plans can be established to maintain time, quality, risk, and scope objectives. Costs, however, are nearly always negatively affected as schedules slip or scope increases. In addition, if unforeseen costs are incurred, adjusting or absorbing these new costs to stay on budget is extremely difficult.

In order to counteract these forces, it is important to ensure that any changes that impact project costs are approved, that these changes are identified quickly, and that the changes are controlled as they occur.

The *PMBOK® Guide* advocates the use of **earned value management** (EVM) to integrate, scope, schedule, and resources. EVM promotes common understanding by using common metrics, and it helps assess the magnitude of cost variations quickly. It also accurately reflects the variations associated with schedule and costs separately. EVM provides a consistent methodology for measuring performance across projects.

Crashing with Time and/or Cost Tradeoffs

In Chapter 6 on time management, crashing was discussed as a strategy to compress the project schedule without reducing project scope. Crashing requires using alternative strategies for completing project activities (such as using outside resources) for the least additional cost. Figure 7-3 below recalls the schedule compression table in Chapter 6, Figure 6-9, but with the cost component added.

Figure 7-3 Schedule Compression with Cost Component

Possible Opportunity	Schedule	Cost	Risk
Reduce Activity W 3 days by hiring an expert	Reduce 3 days	Increase $2,000 for the 3 days	Low
Break Activity W into 2 equal activities and hire another resource at the current rate	Reduce 3 days	Increase $600 for the 3 days	High
Reduce Activity T 2 days by working over the weekend	Reduce 2 days	Increase $400 for the 3 days	Medium

In order to control crashing costs, the following steps should be followed:
- Isolate the critical path; crashing should be performed on activities on the critical path
- Calculate costs for the estimated duration of each activity
- Calculate crash cost per time unit for each activity
- Begin with those activities in which the crash cost per time unit saved is the lowest
- Continue with the next lowest crash cost per time unit saved activity until the desired reduction in project schedule is achieved
- Note that crashing the critical path may result in additional or new critical paths in which additional crashing of activities must occur

Earned Value Terms and Formulas

Some of these are the same terms and formulas used in time management, but in cost management, there are additional terms that facilitate cost management and forecasting.

Three dimensions of EVM are planned value, actual cost, and earned value. These are the earned value terms:
- **Planned value** (PV) is the sum of the approved cost estimates for activities scheduled to be performed during a given period
- **Actual cost** (AC) is the amount of money actually spent in completing work in a given period
- **Earned value** (EV) is the sum of the approved cost estimates for activities completed during a given period

There are many techniques to measuring earned value, which represents work accomplished. A simple percent complete formula may not be determined accurately, so other techniques may be more appropriate. Some techniques to use with non-recurring tasks are:
- **Milestones with weighted values**: specific milestones are defined, and each milestone is given a specific value that will be earned upon completion; this value is used for longer activities that exceed normal review points

7

- **Fixed formula**: a percentage of the project is earned when the activity starts and the remaining percentage is earned when the activity is complete; for example, 25%-75% or 50%-50%; this is used for shorter work packages
- **Percent complete**: a subjective estimate of the percentage of work that has been earned for the given time frame; this is appropriately used for longer activities when work packages are specific and measurable
- **Percent complete with milestone gates**: a subjective estimate of the percentage of work that has been earned up to a milestone gate; additional value cannot be earned until the activity is completed; this is used for longer activities

Variables monitored are cost and schedule variances, the cost and schedule performance indices, and budget and estimate at completion. These are the earned value formulas:

- **Cost variance** (CV) is earned value (EV) minus actual cost (AC); it is the difference between the budgeted cost of the work completed and the actual cost of completing the work; a negative number means the project is over budget
  - $CV = EV - AC$
- **Schedule variance** (SV) is earned value (EV) minus planned value (PV); it represents the difference between what was accomplished and what was scheduled; a negative number means the project is behind schedule
  - $SV = EV - PV$
- **Cost performance index** (CPI) is earned value (EV) divided by actual cost (AC); it is the ratio of what was completed to what it cost to complete it; values less than 1.0 indicate we are getting less than a dollar's worth of value for each dollar we have actually spent; CPI measures cost efficiency
  - $CPI = EV \div AC$

EXAM TIP

According to Fleming and Koppelman in *Earned Value Project Management* (pages 23 to 24), CPI and SPI may be used to determine the efficiency of a project when the project is at least 20% complete. That is, a project that is 10% over budget after 20% of the work is complete will probably overrun the entire project budget by at least 10%.

- **Schedule performance index** (SPI) is earned value (EV) divided by planned value (PV); it is the ratio of what was actually completed to what was scheduled to be completed in a given period; values less than 1.0 mean the project is receiving less than a dollar's worth of work for each dollar we were scheduled to spend; SPI measures schedule efficiency
  - SPI = EV ÷ PV

Forecasting

There are a number of EVM techniques to use for forecasting the remaining work, or estimate to complete, to arrive at the estimate at completion.

- **Budget at completion** (BAC) is the estimated total cost of the project when completed and could be considered the project baseline; **percent spent** is the ratio of actual cost to project budget:
  - AC ÷ BAC
- **Estimate to complete** (ETC) is the estimated additional costs to complete activities or the project
- **Estimate at completion** (EAC) is the amount we expect the total project to cost on completion and as of the "data date" (time now); there are four methods listed in the *PMBOK® Guide* for computing EAC; three of these methods use a formula to calculate EAC; each of these starts with AC or actual costs to date and uses a different technique to estimate the work remaining to be completed, or ETC; the question of which to use depends on the individual situation and the credibility of the actual work performed compared to the budget up to that point
  - A **new estimate** is most applicable when the actual performance to date shows that the original estimates were fundamentally flawed or when they are no longer accurate because of changes in conditions relating to the project
    EAC = AC + New Estimate
    for Remaining Work

- The **original estimate** formula is most applicable when actual variances to date are seen as being the exception and the expectations for the future are that the original estimates are more reliable than the actual work effort efficiency to date

  $$EAC = AC + (BAC - EV)$$

- The **performance estimate low** formula is most applicable when future variances are projected to approximate the same level as current variances

  $$EAC = AC + [(BAC - EV) \div CPI]$$

  A shortcut version of this formula is

  $$EAC = BAC \div CPI$$

- The **performance estimate high** formula is used when a project is over budget and the schedule impacts the work remaining to be completed

  $$EAC = AC + [(BAC - EV) \div (CPI)(SPI)]$$

- **Variance at completion** (VAC) is the difference between the total amount the project was supposed to cost (BAC) and the amount the project is now expected to cost (EAC)
  - $VAC = BAC - EAC$

- **To complete performance index** (TCPI) determines the cost performance that must be attained to meet the BAC or EAC; based on the BAC, the formula is
  - $TCPI = (BAC - EV) \div (BAC - AC)$

  If the EAC has been approved, it will be used instead of the BAC; the formula based on the EAC is
  - $TCPI = (BAC - EV) \div (EAC - AC)$

Figure 7-4 on the following page demonstrates several of the earned value terms and calculations on a project **S-curve**.

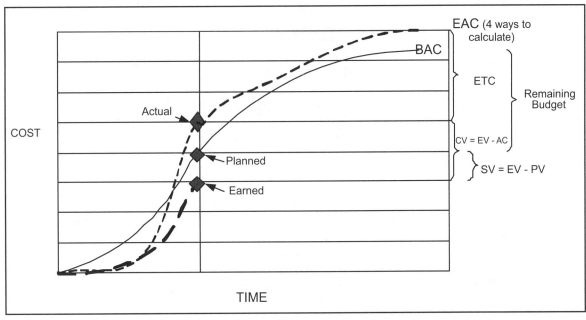

Figure 7-4
Earned Value
S-Curve

Case Study Exercise

Exercise 7-2: After the garage is framed, the roof installed, and the plumbing, electrical, and HVAC are in, the owner asks for a formal project review. You are tasked with creating the earned value reports for the Lawrence Garage Project based on the following actuals-to-date data below.

Phase	Actual Start Variance	Actual End Variance	Actual Duration Variance	Actual Resources Variance	Actual Materials Costs Variance	Notes
Site Prep	+4	+4	Per plan	Per plan	Per plan	Permit delay
Foundation	+6	+7	+1	Per plan	Per plan	Rain delay
Framing	+8	+9	+1	Per plan	Per plan	Materials delay
Utilities	+12	+12	Per plan	Per plan	+8%	Materials delay
Roof	+13	+12	−1	Per plan	Per plan	Materials delay
Interior						
Exterior						
Grounds						
Acceptance						

Profitability Measures

Many projects require an evaluation of the projected profits or returns for the dollars expended. There are a variety of ways to measure profit:

- **Return on sales** (ROS) measures the ratio of profit to total sales
  - ROS = gross profit ÷ total sales or
  - ROS = net profit ÷ total sales
- **Return on investment** (ROI) measures the ratio of profit to total investment, or, if it measures the ratio of profit to total assets, it is called **return on assets** (ROA)
  - ROI = net profit ÷ total investment
  - ROA = net profit ÷ total assets
- **Present value** (PV) (not to be confused with planned value) is the value today of future cash flows, based on the concept that payment today is worth more than payment in the future because we can invest the money today and earn interest on it; if M = amount of payment t years from now and R is the interest rate (also known as the discount rate), then
  - $PV = M \div (1 + R)^t$

  For example, if we think we can get 2% interest for the next 2 years, how much must we invest now in order to have $1,000 2 years from now?
  - $PV = 1,000 \div (1 + 0.02)^2$
    $= 1,000 \div 1.0404$
    $= 961.17$
    - For **net present value** (NPV) (as of a cash flow), add up the PVs over the number of years
- **Internal rate of return** (IRR) is the percentage rate that makes the present value of costs equal to the present value of benefits
- **Benefit cost ratio** (BCR) is just that, a ratio of benefits to costs. For example, an organization may establish that the BCR should be >1.3 before considering a project
  - BCR = PV(revenue) ÷ PV(cost)
- **Payback method** is the amount of time it takes to earn back investment costs; organizations set criteria for acceptable payback periods

EXAM TIP

When you see PV in a question, make sure you understand the context. Is the situation about PV as it relates to estimates (planned value) or PV as it relates to return on investment (present value)?

Notes:

SAMPLE PMP EXAM QUESTIONS ON COST MANAGEMENT

1. Your project data shows that, at some point in the time during execution, the earned value (EV) was $10,000 and the actual cost (AC) was $7,500. The cost variance (CV) at that point was:

   a) There is insufficient data to make a determination
   b) −$2,500
   c) 1.333
   d) $2,500

2. The time-phased budget that is used as a basis for performance measurement is called a:

   a) Schedule baseline
   b) Cost baseline
   c) Gantt chart
   d) Cash flow plan

Software Development Project			
Task	PV	AC	EV
Requirements	200	150	200
Design	500	540	490
Development	850	750	250
Testing	400	300	200
Training	150	150	100

3. In the above table for a software development project, which task has been completed?

   a) Requirements
   b) Development
   c) Testing
   d) Training

✱4. You are the project manager on a construction project that is 50% complete. At this point, the CPI is 1.12. Total earned value to date is $6,300,000, and the original budget was $12,600,000. What is the actual cost?

a) $6,300,000
b) $12,600,000
c) $7,056,000
d) $5,625,000

5. Which of the following elements of the project management plan are inputs to the Plan Cost Management process?

a) Organizational process assets
b) Enterprise environmental factors
c) Scope and schedule baselines
d) Project charter

6. As the project manager assigned to improve a strategic business process, you have been given a budget of $3.0 million to deliver within 12 months. The project is a key strategic initiative this year, and if delivered, the organization will experience a 30% reduction in process costs. Your boss asks you to evaluate the budget and scope and validate the expected ROI of the project. Your experience and knowledge of the costs from prior similar projects you have worked makes you believe that the project costs will exceed the plan by at least 30%. What type of estimating technique is this?

a) Bottom-up
b) Expert judgment
c) Analogous
d) Deterministic

Notes:

7. Of the four tools and techniques for the Control Costs process, which one integrates cost and schedule information as a key element of its approach?

a) Performance reviews
b) Forecasting
c) To-complete performance index
d) Earned value management

8. After the 5th month on her project, a project manager found that the cumulative actual expenditures totaled $120,000. The planned expenditures for this length of time were $100,000. The work completed to date has been budgeted for $102,000. How is the project doing?

a) It is over budget and behind schedule
b) It is in trouble
c) It is over budget and ahead of schedule
d) It is under budget and ahead of schedule

9. A project manager reassesses the estimate at completion for his project. Calculating the EAC by adding the remaining project budget (modified by a performance factor) to the actual cost to date is used most often when the:

a) Current variances are viewed as atypical ones
b) Original estimating assumptions are no longer reliable because conditions are changing
c) Current variances are viewed as typical of future variances
d) Original estimating assumptions are considered to be fundamentally flawed

Notes:

10. Cost management is primarily concerned with:

    a) The cost of resources needed to complete schedule activities
    b) The cost of resources for acquiring the project team
    c) The cost of resources needed to complete the schedule activities, excluding subcontracted items.
    d) The cost of resources and materials for on-going operations

11. As the project manager on a medical device project, you are asked to put together a budget for the project. All of the following are factors you should consider EXCEPT:

    a) Work package cost estimates and how often you will have to report on project costs
    b) The Gantt chart and resource calendars showing when resources will be working on the project
    c) Agreements signed for contract labor and costs associated with risk responses
    d) Scope, cost, and schedule baselines

12. The project budget should include all costs for:

    a) All resources that will be charged to the project
    b) The project and related projects in the program
    c) Contingencies for unknown risks
    d) Ongoing operation and maintenance of the project's product

13. Emma is estimating project costs on a new product development project and has obtained single point estimates from her team members for each activity. What could she use to improve the accuracy of the estimate?

    a) Analogous estimates
    b) Parametric estimates
    c) Bottom-up estimates
    d) Three-point estimates

Notes:

14. The WBS component which is used for project cost accounting is called a/an:

   a) Cost account
   b) Work package
   c) Organizational Breakdown Structure (OBS)
   d) Control account

15. The use of parametric estimating is most reliable when:

   a) Performed on each work package and summed
   b) You have a significant amount of relevant historical data
   c) All risks have response strategies defined
   d) Using the experience of other similar projects

Notes:

7

ANSWERS AND REFERENCES FOR SAMPLE PMP EXAM QUESTIONS ON COST MANAGEMENT
Section numbers refer to the *PMBOK® Guide*.

1. **D** **Section 7.4.2.1 – Monitoring and Controlling**
   $CV = EV - AC$; $CV = \$10,000 - \$7,500 = \$2,500$; when doing the calculations, always start with EV; C) is the CPI.

2. **B** **Section 7.3.3.1 – Planning**
   A) is not a budget output; C) a Gantt chart is a used to measure schedule performance, but not cost; D) would be a time-phased budget, but it is not used as a basis for performance measurement.

3. **A** **Section 7.4.2.1 – Monitoring and Controlling**
   For requirements, the $SV = EV - PV = 2.00 - 2.00 = 0$. Zero SV means that a task is complete.

4. **D** **Section 7.4.2.1 – Monitoring and Controlling**
   $CPI = EV \div AC$, so $AC = EV \div CPI$; if $CPI = 1.12$, then $AC = \$6,300,000 \div 1.12 = \$5,625,000$.

5. **C** **Section 7.1.1 – Planning**
   A), B), and D) are not elements of the project management plan.

6. **C** **Section 7.2.2.2 – Planning**
   Since the estimate is based on your experience and knowledge of prior similar projects, the correct answer is analogous.

7. **D** **Section 7.4.2.1 – Monitoring and Controlling**
   Earned value management integrates schedule and cost performance with the baseline plan and actual costs.

8. **C** **Section 7.4.2.1 – Monitoring and Controlling**
   $EV = \$102,000$
   $PV = \$100,000$
   $AC = \$120,000$
   $SV = \$102,000 - \$100,000 = \$2,000$
   $CV = \$102,000 - \$120,000 = <\$18,000>$

9. C Chapter 7.4.2.1 – Monitoring and Controlling
Since the project manager is using a remaining budget modified by a performance factor, he or she is assuming future work will reflect current performance.

10. A Section 7.0 – Planning
B) resources are one component of cost; C) subcontracted items should be included in both estimates and budgets; D) ongoing operations are not part of the project cost.

11. D Section 7.3.1 – Planning
The output of the Determine Budget process is the cost baseline.

12. A Section 7.3 – Planning
B) related projects will have their own project budgets; C) management reserves are separate from the project budget; D) ongoing maintenance or operation is not included in the project budget since the project is finite.

13. D Section 7.2.2.5 – Planning
A) analogous estimates are less time consuming but typically less accurate; B) parametric estimates can be very accurate, but it's not likely this project will have adequate historical information to use for parametric estimating; C) the team members have prepared a bottom-up estimate since they have provided estimates for each activity.

14. D Section 7.1.3.1 – Planning
A) is an accounting term; B) is the lowest level of the WBS; C) the OBS identifies an individual's reporting relationship to the work packages.

15. B Section 7.2.2.3 – Planning
The more data provided to parametric models, theoretically, the more accurate the model will be. A) summing each work package is bottom-up estimating; C) the completion of the Plan Risk Responses process is not required to produce a parametric estimate; D) is the definition of analogous estimating.

CASE STUDY SUGGESTED SOLUTIONS

Exercise 7-1
S-Curve for the Lawrence Garage Project

The S-curve below was generated using the data at the bottom of this page:

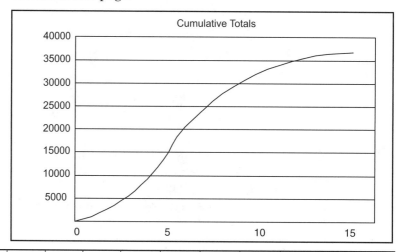

Week Phase	1	2	3	4	5	6	7	8	9	10	11	12	13
PM	384	384	384	384	384	384	384	384	384	384	384	384	384
Site	1,000												
Foundation	800	1,700											
Framing		7,000	1,000	1,000	1,000								
Utilities			500	1,000	2,500								
Roof						4,500							
Stucco							2,300	2,200					
Insulation									550				
Drywall									1,500	800			
Paint											500		
Interior												600	
Grounds								300	1,200				
Acceptance													225
Weekly Totals	2,184	9,084	1,884	2,384	3,884	4,884	2,684	2,884	3,634	1,184	884	984	609
Cumulative Totals	2,184	11,268	13,152	15,536	19,420	24,304	26,988	29,872	33,506	34,690	35,574	36,558	37,167

Exercise 7-2
Earned Value Reports for the Lawrence Garage Project

Using the information in this exercise, we extrapolated the following data:

Week Phase	1	2	3	4	5	6	7	8	9	10	11	12	13
PM	384	384	384	384	384	384	384	384	384	384	384	384	384
Site		1,000											
Foundation		800	1,500	200									
Framing				7,000	1,000	1,000							
Utilities					500	1,150	2,500						
Roof								4,500					
Stucco							2,300	2,200					
Insulation									550				
Drywall									1,500	800			
Paint										500			
Interior												600	
Grounds								300	1,200				
Acceptance													225
Weekly Totals	384	2,184	1,884	7,584	1,884	2,534	5,184	7,384	3,634	1184	884	984	609
Actuals	*384*	*2,184*	*1,884*	*7,584*	*1,884*	*2,534*	*2,884*	*4,884*					
Cumulative Actuals	384	2,568	4,452	12,036	13,920	16,454	19,338	24,222	24,222	24,222	24,222	24,222	24,222

These data yielded the results on the following page, based on the cumulative numbers and the planned numbers from Exercise 7-1.

Week	1	2	3	4	5	6	7	8	9	10	11	12	13
Cum EVT													
PV	2,184	11,268	13,152	15,536	19,420	24,304	26,988	29,872	33,506	34,690	35,574	36,558	37,167
EV	384	2,568	4,452	12,036	13,920	16,304	19,188	24,072					
AC	384	2,568	4,452	12,036	13,920	16,454	19,338	24,222					
SV	−1,800	−8,700	−8,700	−3,500	−5,500	−8,000	−7,800	−5,800					
SVI	0.176	0.228	0.339	0.775	0.717	0.671	0.711	0.806					
CV	0	0	0	0	0	−150	−150	−150					
CVI	1.000	1.000	1.000	1.000	1.000	0.991	0.992	0.994					

Looking at the schedule variance (SV), we can see that we fell behind right away in the first week. We fell even more in the second week, but then picked up after the third week. However, we fell behind again before finally improving over the past two weeks. According to the SVI, then, we are currently at about 80% of where we had planned to be at this time.

Cost variance (CV) is better. In fact, for the tasks we completed, we were right on target for the first five weeks. In week six we went over budget, but have not yet had any additional budget slippage. The CVI shows that we are getting about 99 cents worth of value for every dollar spent.

QUALITY

CHAPTER 8 | **QUALITY**

8

8

QUALITY MANAGEMENT

The quality management questions on the PMP exam are straightforward—especially if you know the definitions of terms and understand statistical process control. You are not required to solve quantitative problems, but there are questions on statistical methods of measuring and controlling quality.

An emphasis on customer satisfaction, cost of quality, management's responsibility for quality, and continuous improvement is likely to be on the exam; how tools such as Pareto and cause-and-effect diagrams are used may also be on the exam.

To pass, you must know the differences among the three quality processes: Plan Quality Management, Perform Quality Assurance, and Control Quality.

Many organizations use these terms interchangeably; however, the *PMBOK® Guide* specifically defines each in terms of the process group in which it is performed. The following chart summarizes the three processes as defined in the *PMBOK® Guide* and questioned within the exam.

Process	Plan Quality Management	Perform Quality Assurance	Control Quality
Primary Activity	Plan	Implement/Manage	Measure/Monitor
Explanation	Determine what the quality standards for the project will be and document how the project will be measured for compliance.	Use the measurements to see if the quality standards will be met; validation of the quality standards.	Perform the measurements and compare them to specific quality standards; identify ways of eliminating problems in the future.
Process Group	Planning	Executing	Monitoring and Controlling

Things to Know

1. The three processes of quality management:
   - **Plan Quality Management**
   - **Perform Quality Assurance**
   - **Control Quality**
2. Quality Management General Concepts
   - **Quality** versus **grade**
   - **Precision** versus **accuracy**
   - The **impact of poor quality**
   - The **legal implications of quality**
   - **Market expectation**s regarding quality
   - Where **responsibility for quality** lies
   - **Prevention** over **inspection**
   - **Quality policy**
   - **Quality objectives**
3. **Quality theories**
   - **Deming**
   - **Crosby**
   - **Juran**
   - **Taguchi**
4. **Quality approaches**
   - **Total quality management** (TQM)
   - **Continuous improvement process** (CIP)
   - **Just in time** (JIT) and the **six sigma initiatives**
5. The concept of **cost of quality** (COQ)
6. The seven basic tools of quality management:
   - **Cause and effect diagrams**
   - **Flowcharts**
   - **Checksheets**
   - **Pareto diagrams**
   - **Histograms**
   - **Control charts**
   - **Scatter diagrams**
7. The concept of **statistical sampling**
8. The importance of the **process improvement plan**
9. The contents of the **quality management plan**
10. The **quality audit**
11. The differences between the **Control Quality** and **Validate Scope** processes
12. Key **interpersonal skills** for success

Key Definitions

Accuracy: the assessment of correctness.

Benchmark: comparing actual or planned project practices to those of comparable projects to identify best practices and generate ideas for improvement. Benchmarks provide a basis for measuring performance.

Capability maturity model integration (CMMI): defines the essential elements of effective processes. It is a model that can be used to set process improvement goals and provide guidance for quality processes.

Design of experiments: a statistical method for identifying which factors may influence specific variables of a product or process either under development or in production.

Grade: the category or level of the characteristics of a product or service.

Lean Six Sigma: a business improvement methodology that strives to eliminate non-value added activities and waste from processes and products.

Malcolm Baldrige: the national quality award given by the United States' National Institute of Standards and Technology. Established in 1987, the program recognizes quality in business and other sectors. It was inspired by Total Quality Management.

Organizational project management maturity model (OPM3®): focuses on the organization's knowledge, assessment, and improvement elements.

Precision: a measure of exactness.

Process quality: specific to the type of product or service being produced and the customer expectations, the level of process quality will vary. Organizations strive to have efficient and effective processes in support of the product quality expected. For example, the processes associated with building a low-quality, low-cost automobile can be just as efficient, if not more so, than the processes associated with building a high-quality, high-cost automobile.

Product quality: specific to the type of product produced and the customer requirements, product quality measures the extent to which the end product(s) of a project meets the specified requirements. Product quality can be expressed in terms that include, but are not limited to, performance, grade, durability, support of existing processes, defects, and errors.

Project quality: typically defined within the project charter, project quality is usually expressed in terms of meeting stated schedule, cost, and scope objectives. Project quality can also be addressed in terms of meeting business objectives that have been specified in the charter. Solving the business problems for which the project was initiated is a measure of the quality for the project.

Quality: the degree to which a set of inherent characteristics satisfies the stated or implied needs of the customer. To measure quality successfully, it is necessary to turn implied needs into stated needs via project scope management.

Quality objective: a statement of desired results to be achieved within a specified time frame.

Quality policy: a statement of principles for what the organization defines as quality.

Six Sigma: an organized process that utilizes quality management for problem resolution and process improvement. It seeks to identify and remove the causes of defects.

Statistical sampling: involves choosing part of a population of interest for inspection.

Warranties: assurance that the products are fit for use or the customer receives compensation. Warranties could cover downtime and maintenance costs.

Quality Management General Concepts

Quality and **grade** are different. Grade is a way to distinguish between products with the same functional use but different technical attributes. For example, a grade one bolt has a certain strength while a grade three bolt of the same size is stronger, and a grade five bolt is stronger still. The grade one bolt, though of low grade, can still be of high quality (no defects, of proper size, etc.).

Precision and **accuracy** are also different. Precision demonstrates a consistency that can be counted on, while accuracy measures whether the results are what were expected or desired. Let's look at precision and accuracy in a GPS system. If you enter in a particular address and you ALWAYS arrive at the same location, but that location is not correct, that would be precision. If you ALWAYS arrive at the correct location whenever you use your GPS, that is accuracy.

Quality, grade, precision, and accuracy are important to consider and need to be defined at appropriate levels for the project within the **quality management plan**.

Regardless of the organization, the **impact of poor quality** can be significant; it can result in higher costs to the entity or the customer, less customer satisfaction, lower team morale, and greater risk of project failure.

In addition to the benefits organizations gain from implementing quality programs, there are **legal implications of quality** that must be addressed when developing the quality management plan:
- Criminal liability
- Fraud or gross negligence
- Civil liability
- Criminal or civil liability, even if following orders
- Lawsuits against the company
- Appropriate corporate actions

In determining what preventative measures to take to avoid nonconformance costs, the project manager must take into consideration the **market expectations** of the project's product by reviewing the customer's product expectations on the following criteria:
- Salability: a balance of quality and cost
- Producibility or constructibility: the ability of the product to be produced with available technology and workers at an acceptable cost
- Social acceptability: the degree of conflict between the product or its process and the values of society
- Operability: the degree to which a product can be safely operated
- Availability: the probability that the product, when used under given conditions, will perform satisfactorily; the two key parts of availability are:
  - Reliability: the probability that the product will perform, without failure, under given conditions for a set period of time
  - Maintainability: the ability of the product to be restored to its stated performance level within a specified period of time

EXAM TIP
Be sure to read the exam questions on responsibility for quality carefully in order to determine to whom in the organization the question refers.

The organization as a whole has a **responsibility for quality**, and what is recommended in the *PMBOK*® *Guide* may not be what your organization practices. The position followed by the *PMBOK*® *Guide* is:

- The project manager has the ultimate responsibility for the quality of the product of the project (in reality, the project manager may delegate work but must retain responsibility)
- The team member has the primary responsibility for quality at the task or work package level
- The primary responsibility for establishing design and test criteria resides with the quality engineer

Whoever is responsible for quality within an area of expertise must identify quality problems, recommend solutions when problems occur, implement solutions, and, if the process is nonconforming, limit further processing.

For many years, the determination of quality relied heavily on **inspection** methods. Over the years, it has been determined that the costs of inspection can become so high that it is better to spend money on **prevention** to keep problems from ever occurring. The *PMBOK® Guide* supports the notion that quality must be planned in and not inspected in. In reaching this conclusion, PMI researched the work of many key quality experts from the past several decades. The exam could include questions on specific theories or experts' opinions. See the quality theories section in this chapter for a deeper discussion of these theories and opinions.

> **EXAM TIP**
> The *PMBOK® Guide* emphasizes that quality should be planned into the project, not inspected in.

8

Organizations may have a statement defining the principles for quality within the organization. This is generally known as the organization's **quality policy**.

It is not recommended to develop a quality policy from scratch. Most organizations have a quality policy endorsed by senior management which can be adopted or adjusted to fit the needs of a project. However, if a project spans more than one key organization in a joint partnership, a project quality policy may need to be developed.

The quality policy does NOT define how quality will be achieved. When organizations create quality policies, they do so to promote consistency, to provide specific guidelines for important matters, and to help outsiders better understand the organization. For example, some companies consider quality as the ability to produce products very inexpensively and want to be considered the low-cost leader, while others prefer to offer the most options or features for a higher price. Successful quality policies are drafted by specialists, approved by top management, and understood by and adhered to by the entire organization.

Quality objectives are written for a project by the project team. In writing quality objectives, it is important to realize the perspective from which they are being written. For customers, quality is typically defined by the ability of the project's product to be **fit-for-use**. From a project perspective, adherence to specifications will define quality. Whether defining objectives from the customer's or the project's perspective, it is important to define goals specifically through stated quality objectives and ensure that they are communicated well and understood by all stakeholders.

Quality Theories

There are four well-known quality theories that you may see questions on in the exam.

W. Edwards **Deming** is well known for his four-step cycle to improve quality: **plan-do-check-act** (PDCA). He also developed 14 activities for implementing quality. For the exam, know some major points of his works:
- Use a participative approach to quality
- Adopt a new philosophy of quality throughout the organization
- Cease the use of mass inspections
- End awards based on price
- Improve production and service
- Institute leadership
- Eliminate numerical quotas
- Emphasize education and training
- Encourage craftsmanship

> **EXAM TIP**
> Know the differences between the various quality theories defined in the *PMBOK®* *Guide* Figure 8-2.

Philip **Crosby** is also well known for his books on quality. Similar to Deming, he too developed 14 steps to improving quality. These steps emphasize management commitment, measurement, zero defect planning, goal setting, quality awareness, and quality councils.

In addition, Crosby stressed four absolutes of quality:
- Quality is conformance to requirements
- The system of quality is prevention
- The performance standard is zero defects
- The measure of quality is the price of nonconformance

Joseph **Juran** developed the **fit-for-use** concept of quality which emphasizes that the measure of high quality is achieved by ensuring that the product meets the expectations of the stakeholders and customers.

Juran's fitness-for-use concept looks at three components of quality. These components are known as the Juran trilogy:
- **Quality of design**: design may have many grades
- **Quality of conformance**: determined by choice of process, training, adherence to program, and motivation
- **Quality characteristics**: determine the characteristics important to the customer
  - Structural (length and frequency)
  - Sensory (taste and beauty)
  - Time oriented (reliability and maintainability)
  - Ethical (courtesy and honesty)

Juran also established the following trilogy as an approach to improving quality:
- **Plan**: attitude breakthrough, identify vital few new projects
- **Improve**: knowledge breakthrough, conduct analysis, institute change
- **Control**: overcome resistance, institute controls

Dr. Genichi **Taguchi** developed the concept of the **loss function** according to which, as variation for the target increases, losses will also increase. Taguchi's rule for manufacturing is based on the concept that the best opportunity to eliminate variation is during the design of a product and its manufacturing process. The Taguchi loss function can be used to measure financial loss to society resulting from the poor quality of a product.

Quality Approaches

There are many non-proprietary approaches to quality such as total quality management and continuous improvement, which have been popular in Japan, and Six Sigma and Lean Six Sigma, which are more popular in the United States.

Although there is no one definition of **total quality management** (TQM), most definitions include providing quality products at the right time and at the right place, thereby meeting or exceeding customer requirements. Harold Kerzner, in *Project Management: A Systems Approach to Planning, Scheduling, and Controlling*, has defined seven primary strategies for TQM (pages 806 to 809):

- Solicit improvement ideas from employees
- Encourage teams to identify and solve problems
- Encourage team development
- Benchmark every major activity in the organization
- Utilize process management techniques
- Develop staff to be entrepreneurial and innovative in dealing with customers and suppliers
- Implement improvements in order to qualify for ISO 9000

The **continuous improvement process** (CIP) or **Kaizan** is another approach to quality. Kaizan is the Japanese word for a sustained gradual change for improvement. It differs from **innovation**, which consists of sudden jumps that plateau and mature over time before the next jump. The **plan-do-check-act** cycle developed by Deming is the basis for CIP. The Japanese also came up with the concept of providing materials only when they are needed in manufacturing environments. This concept is known as **just in time** (JIT). TQM and **Six Sigma initiatives** help to improve project management processes as well as project management products. For more information on Six Sigma, you can go to www.asq.org.

PLAN QUALITY MANAGEMENT PROCESS

The Plan Quality Management process is performed in conjunction with all other project planning. The Plan Quality Management process includes:
- Identifying which **quality standards** are relevant to the project and product and determining how to satisfy them
- **Benchmarking** past projects to find ideas for improvements and to establish quality performance measures
- Using **cost benefit analysis** to compare the benefits and the costs of quality
- Flowcharting a process or system to show how various components interrelate (used to help determine potential future quality problems and establish quality standards)
- Having a **design of experiments** with **"what if" scenarios** to determine which variables will have the most influence on project outcomes, thereby improving quality

Cost of Quality (COQ)

Conformance is the ability for the product of the project to meet requirements. A project manager has options when planning a project. He or she can implement quality processes to increase the likelihood that the products will meet requirements, or the project manager can inspect the product, determine if it meets requirements, and take corrective action if it does not.

The *PMBOK® Guide* advocates a Deming approach. Deming's approach says that approximately 85% of the costs of quality are the direct responsibility of management. These costs can be broken up into two categories: the costs of conformance and the costs of nonconformance.

Costs of Conformance	Costs of Nonconformance
Quality procedures	Scrap
Quality training	Rework
Studies	Warranty costs
Surveys	Inventory costs
Validation and audits	

Costs of conformance can be categorized as **prevention costs**, such as the use of high-quality parts and documented processes, and **appraisal costs**, which assess quality at various stages of the project to increase the likelihood of conformance.

Costs of nonconformance can be categorized as internal or external failures.

- **Internal failures** are the costs associated with scrapping or reworking a product before it reaches the end customer
- **External failures** are those that have reached the customer; external failures include costs associated with handling and resolving customer concerns

In order to implement a quality plan successfully, training and coaching may be needed in order to keep quality at the top of the priority list.

Seven Basic Quality Tools

There are many exam questions on quality tools. You will need to know the uses and differences in the following tools and techniques. In general, to use these control tools effectively, an agreement must be made on what will be observed, the time frame for observation, the form of the presentation, and how the data will be collected.

Cause and effect diagrams (also called fishbone or Ishikawa diagrams) show how various causes and subcauses relate to create problems or effects.

Flowcharts show how various elements of a system relate. System or process flowcharts are the most common types of flowcharts.

Issue	Week 1	Week 2	Week 3	Week 4	Total
One	III	I	III	I	8
Two	II	IIII	I	II	9
Three	IIII	III	II	II	11
Total	9	8	6	5	28

Figure 8-1
Sample Checksheet

Checksheets are a simple way to gather data. Checksheets are orderly and help ensure that all components needing to be reviewed are reviewed and that key components are not overlooked or excluded. A sample checksheet is shown in Figure 8-1 above.

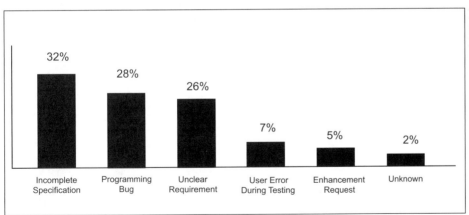

Figure 8-2
Sample Pareto Diagram

Pareto Diagrams are histograms ordered by frequency of occurrence. Pareto diagrams are conceptually related to Pareto's law, which visually shows that 20% of causes produce 80% of defects. Figure 8-2 above shows a sample Pareto diagram.

Histograms or **vertical bar charts** are commonly used in statistics as a graphical display of tabulated frequencies. The categories are usually denoted on the x-axis with the height of the bar displaying the proportion of cases that fall into each category. See Figure 8-3 below for a sample histogram.

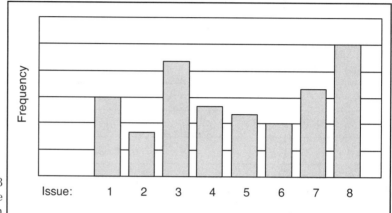

Figure 8-3 Sample Histogram

Control charts give a graphical display of results of a process over time. Figure 8-4 below shows a sample control chart. Control charts include a defined upper and lower control limit, a mean, and a visual pattern indicating out-of-control conditions, such as **outliers** (points outside upper [UCL] or lower [LCL] control limits).

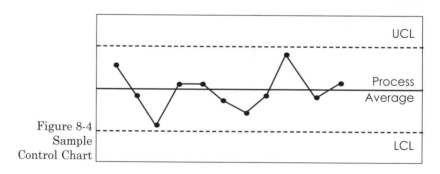

Figure 8-4 Sample Control Chart

Control charts that produce particular patterns can provide visual information to the project manager. Some such patterns are:

- **Limit huggers**: a run of points close to control limits
- **Run**: a series of consecutive points on the same side of the mean
- **Trend**: a series of consecutive points with an increasing or decreasing pattern
- **Cycle**: a repeating pattern of points
- **Rule of seven**: a run of seven or more points above or below the mean indicating adjustment is needed

Scatter diagrams are used to show the correlation between two characteristics. If there is a strong correlation, minor changes to one variable will change the other variable. The relative correlation of one characteristic to the other can be seen by the pattern formed by the clusters of dots in the scatter diagram. If the cluster approaches a line in appearance, then the two characteristics are said to be linearly correlated. Figure 8-5 below shows a sample scatter diagram.

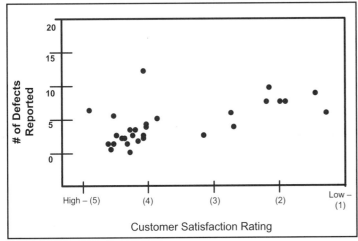

Figure 8-5 Sample Scatter Diagram

Statistical Sampling

Statistical sampling involves choosing part of a population for inspection for the purpose of accepting or rejecting the entire lot. The results of statistical sampling can be depicted through the use of a variety of charting methods such as histograms, scatter diagrams, or Pareto diagrams.

The advantages of using sampling techniques include less product damage, the ability to make decisions more quickly, fewer expenses, and fewer resources. The disadvantage is the possibility of bad decisions due to incomplete information. Here are some sampling definitions:

- **Attribute**: characteristic of the product that is appraised in terms of whether or not it exists
- **Variable**: anything measured
- **Sampling plan**: must include the sample size and the acceptance criteria
- **Producer's risk**: the chance of rejecting a good lot prior to selling to the customer
- **Consumer's risk**: the chance of accepting a bad lot after purchase

Inspections are used after the work is completed. **Checksheets** or **data tables** may be used to assist in the measuring, examining and testing activities. Figure 8-6 below shows a sample inspections checksheet.

Figure 8-6
Sample
Inspections
Checksheet

Problem	Month			
	1	2	3	Total
A	‖	‖	‖	5
B	‖	‖	‖	3
C	‖‖‖	‖	‖‖‖	12
Total	8	5	7	20

Process Improvement Plan

The *PMBOK® Guide* emphasizes the importance of continually assessing the gap between the current position of the organization and the desired goals or capabilities to be achieved. This continual activity of planning and implementing processes within the organization to improve is known as process improvement. Therefore, process improvement is an analytical approach that focuses on activities that provide value to the organization.

There are many well-known process improvement models available such as **CMMI**, **Malcolm Baldrige**, and **OPM3®**. Each of these models is based on the premise that process improvement is a continuous strategy based on ongoing incremental betterment within an organization. This continuous process improvement provides an iterative means for the ongoing improvement of all processes, specifically the project management and product development process.

The document that details the activities to analyze processes in order to improve value is the **process improvement plan**. This plan should consider:
- Process boundaries
- Process configuration
- Process metrics
- Targets for improved performance

Quality Management Plan

A quality management plan is a key output of the Plan Quality Management process. It should describe how the project management team will implement its quality policy and will provide input to the overall project management plan. A good quality management plan will specifically address each of the following:
- Design control
- Document control
- Purchased material control
- Material identification and control
- **Inspections**

- Test control
- Measuring and testing equipment control
- **Corrective actions**
- Quality assurance records
- **Quality audits**
- **Process improvements**

Therefore, to develop a good quality management plan, the project manager must understand the various tools available and select the appropriate tools that will be needed for the current project.

PERFORM QUALITY ASSURANCE PROCESS

Quality assurance activities occur during the execution phase of the project. It is the process of regular structured reviews to ensure the project complies with the planned quality standards. This is usually done by means of a **quality audit**.

Quality audits are independent evaluations of quality performance to ensure that:
- Intended quality will be met
- Products are safe and fit for use
- Laws and regulations are followed
- Data systems are adequate
- Corrective action is taken, if needed
- Improvement opportunities are identified
- Quality standards, procedures, and methods established during quality planning are reevaluated and are still relevant

A good quality assurance system will:
- Identify objectives and standards
- Be multifunctional and prevention oriented
- Collect and use data
- Establish performance measures
- Include a quality audit

EXAM TIP

Regularly reviewing how quality is being delivered on the project will increase the probability that the project will deliver on the expected business intent. Use of process analysis techniques and best practices should be part of the quality management plan when applicable.

The Perform Quality Assurance process includes quality tools that can be used in addition to the seven basic quality tools defined in the Plan Quality Management process. These additional tools are:
- Affinity diagrams
- Process decision program charts
- Interrelationship digraphs
- Tree diagrams
- Prioritization matrices
- Activity network diagrams
- Matrix diagrams
- Process analyses

Case Study Exercise

Exercise 8-1: Using the data in the table below for the Lawrence Garage Project, create a Pareto diagram for the reasons each phase was completed later than planned. What do you think your results indicate?

Phase	Actual Start Variance	Actual End Variance	Actual Duration Variance	Actual Resources Variance	Actual Materials Cost Variance	Notes
Site Prep & Foundation	4 days delayed	4 days delayed	Per Plan	Per Plan	Per Plan	Permit delay (3 days) & Rain Delay (1 day)
Framing	6 days delayed	7 days delayed	1 day delayed	Per Plan	Per Plan	Lumberyard material delays
Exterior	2 days delayed	4 days delayed	2 days delayed	Per Plan	Per Plan	Paint contractor material delays
Interior	3 days delayed	3 days delayed	Per plan	Per Plan	Per Plan	Lumberyard material delays
Grounds						
Completion						

CONTROL QUALITY PROCESS

The Control Quality process is performed as part of the monitoring and controlling process group. Quality control involves measuring the process or performance using **quality control tools**. It also includes the technical processes that compare and report a project's actual progress with its standard. A good quality control system will:
- Select what to control
- Set standards
- Establish measurement methods
- Compare actuals to standards
- Act when standards are not met

Key outputs of the Control Quality process are the **validated deliverables** and **work performance information**. Based on these recommendations, **corrective actions** are performed as part of the Direct and Manage Project Work process.

Control Quality vs. Validate Scope

The Control Quality process is usually performed before the Validate Scope process because it focuses on:
- The correctness of the deliverables
- Making sure the stated quality requirements have been met

The Validate Scope process focuses on the acceptance of the deliverables by the customer.

EXAM TIP

The project manager must understand the expected business value of the project. If the goal of the project is a 50% return on investment, then the project team must have metrics to identify if the project will be able to meet its stated goal.

EXAM TIP

The Control Quality process is concerned with validating deliverables, NOT receiving acceptance by the client.

KEY INTERPERSONAL SKILLS FOR SUCCESS

The interpersonal skill highlighted in this chapter is:

Coaching

Project managers will be more successful if they help their team members become successful. One way to accomplish this is for the project manager to work with the team members or individuals to help them build new skills or expand existing skills.

In this chapter we discuss many different quality theories, tools, techniques, and approaches. We stated earlier that quality does not just happen because a plan exists. It is up to the project manager and the project team to commit to the level of quality that is expected. The project team members may or may not have experience with each tool and therefore may not be aware that a more appropriate tool exists. The project manager may need to coach the team on the creation and the usefulness of the tools that are chosen in order to ensure the highest quality.

SAMPLE PMP EXAM QUESTIONS ON QUALITY MANAGEMENT

1. You are the project manager for a business process improvement project for a strategic business process that is 50% complete. One component of the existing business process had been targeted for improvement because of the significant spikes in quality problems. The tool you can use to monitor the success of the improvement project that measures compliance within acceptable limits is called a:

   a) Root cause analysis
   b) Quality audit
   c) Control chart
   d) Pareto diagram

2. At a progress meeting, the quality assurance team shows that the number of negative findings per deliverable has increased in the past two weeks. This does not meet the project's expectations. As project manager, the next step you may want to initiate in order to see why the problem is occurring is to:

   a) Develop a flowchart
   b) Call a meeting with the quality team to determine the cause of the problem
   c) Review the schedule variance on the project
   d) Review the vendor contracts to determine if they are at fault

Notes:

3. As a project manager, you find that your customer wants you to develop a medical device product within six months. The requirements are complete and thorough. However, in your assessment after performing a work breakdown and scheduling exercise, you determine that the quality assurance and quality control activities of the project have been underestimated, adding three months to the total project duration. One nonproprietary aspect of quality that you should quickly address with the customer is:

a) Six Sigma
b) Continuous improvement
c) Cost of quality
d) ISO 9000

4. Your project is nearing completion and you have scheduled a deliverable review meeting with your team for next week. The objective of the meeting will be to verify that each project deliverable has met the quality requirements set by the customer. This is an example of the _____ process.

a) Validate Scope
b) Control Quality
c) Inspection
d) Perform Quality Assurance

5. Another term for failure costs is:

a) Sunk costs
b) Opportunity costs
c) Costs of poor quality
d) Punitive damages

Notes:

6. You are a project manager on a product enhancement effort. The current cost of performing all quality activities is planned at $100,000. You anticipate a 20% error rate which costs approximately $10,000 to either fix or scrap. Your engineers have identified a modification to the product that would cost $50,000 to develop, and would drop the error rate down to 10%. You should:

   a) Do nothing because you don't have the budget for the change
   b) Conduct a sales analysis to determine if the cost will be recovered
   c) Proceed with the change since the engineers are highly confident the change will impact the largest issue experienced
   d) Ask the client whether to proceed with the change

7. Your customer's organization has a reputation for providing the highest quality products and is paid well for maintaining that level of quality. As the project manager, you should begin to develop a project management plan and schedule that:

   a) Model your project management plan after prior similar projects
   b) Ensure the appropriate amount of inspection and prevention are incorporated
   c) Perform analogous estimating to create the project budget
   d) Model your project schedule after prior similar projects

8. Standard deviation is a measure of how:

   a) Much time remains in a project
   b) Far you are from the mean
   c) Correct the sample is
   d) Results compare to benchmarks

Notes:

9. As a project manager, one aspect of your responsibility for quality is to ensure the project stakeholders are:

   a) Fully aware of the quality policy
   b) Not included in the development of the quality management plan
   c) Involved in the quality control process
   d) Responsible for managing quality expectations

10. The technique that examines quality practices implemented, nonconformity with quality policies, and compliance with an organization's quality process is:

    a) Quality audits
    b) Control charts
    c) Flowcharting
    d) Process analysis

11. Quality management recognizes the importance of _____.

    a) Tools and techniques
    b) Quality audits
    c) Inputs and outputs
    d) Prevention over inspection

12. The statistical technique that provides a framework for understanding which variables have the most influence on a process is called:

    a) Benchmarking
    b) A flowchart
    c) The SIPOC model
    d) The design of experiments

Notes:

8

13. The project sponsor is not comfortable with the quality level of the project. He instructs the project manager to come up with quality standards and to improve quality. The project manager, however, is concerned about the effect of quality improvements on the project. In the long run, the project should experience:

a) Reduced productivity and an increase in overall product or service cost
b) Reduced productivity and no change to cost effectiveness and cost risk
c) Increased productivity, decreased cost effectiveness, and increased cost risk
d) Increased productivity, increased cost effectiveness, decreased cost risk

14. In order to ensure the customer will accept the project deliverables, it is important that _____ be set early in the project life cycle.

a) Acceptance criteria
b) Roles and responsibilities
c) The change control process
d) The schedule

15. An example of a cost of conformance to quality is:

a) Quality training
b) Scrap
c) Warranty costs
d) Rework

Notes:

8

ANSWERS AND REFERENCES FOR SAMPLE PMP EXAM QUESTIONS ON QUALITY MANAGEMENT

Section numbers refer to the *PMBOK® Guide*.

1. **C Section 8.1.2.3 – Monitoring and Controlling**
   A) is a way to determine the cause, not monitor progress; B) are used to audit the project policies, processes, and procedures, NOT the results of the project; D) identifies defects that are more common than others.

2. **A Section 8.1.2.3 – Monitoring and Controlling**
   Flowcharts can help analyze how problems occur.

3. **C Section 8.1.2.2 – Planning**
   Since this is a medical device, you must look to the customer to determine if the level of quality assurance and control is appropriate for the risk of selling this device to the marketplace.

4. **B Section 8.3 – Monitoring and Controlling**
   A) is the process of obtaining the stakeholders' formal acceptance of the completed project scope and associated deliverables; C) is a technique used within the Control Quality process; D) Perform Quality Assurance makes sure quality processes are followed and does not focus on the quality of each deliverable.

5. **C Section 8.1.2.2 – Planning**
   A) are expended costs; B) are what you give up to pursue a particular course of action; D) are awarded by the courts to punish a defendant who is at fault in a legal case.

6. **B Section 8.1.2.1 – Planning**
   Since it isn't clear that this project is directed at a specific client, you should compare the overall cost of the change to the benefit it will bring the organization.

7. **B Section 8.0 – Planning**
   Although A), C), and D) are things that could be done, the focus must be on providing the highest level of quality for the project, based on the anticipated needs of a known client.

8. **B** **Section 8.1.2.3 – Planning**
Standard deviation measures the amount of variation from the average.

9. **A** **Section 8.1.1 – Planning**
Involving the stakeholders in many aspects of the project is always better than not doing so. How you involve people will be up to you and the situation of the project. At a minimum, you should communicate the quality policy to all stakeholders so that they can perform accordingly.

10. **A** **Section 8.2.2.2 – Executing**
B) control tools such as control charts are used to measure performance; C) flowcharting defines the sequence of activities in a process; D) the objective of process analysis is to identify needed improvements based on problems experienced.

11. **D** **Section 8.0 – Planning**
Both project management and quality management disciplines recognize the importance of prevention over inspection, management responsibility, customer satisfaction, and continuous improvement.

12. **D** **Section 8.1.2.5 – Planning**
A) is used to compare project practices to other comparable projects; B) shows the sequence of steps in a process; C) is a type of flowchart that shows Suppliers, Inputs, Process, Outputs, and Customers.

13. **D** **Section 8.0 – Planning**
In the short run, productivity may decrease, and cost may increase. But quality improvements should result in improvements to the process over time.

14. **A** **Section 8.1.1.1 – Planning**
The success of the project depends on the clarity of acceptance criteria set early on in a project's development. Acceptance criteria should be defined as part of the scope statement and included in the scope baseline.

15. A   Section 8.1.2.2 – Planning
     B), C), and D) are costs of nonconformance.

CASE STUDY SUGGESTION SOLUTION

Exercise 8-1
Pareto Diagrams for the Lawrence Garage Project

Quantity of Delays by Type of Delay

Total Days Delayed by Vendor

Here are two examples of Pareto diagrams that could be created out of the data provided. Looking at the Pareto diagram for the reasons for delays, it appears that our materials suppliers cannot seem to get us what we need when we need it. We would need to investigate further to see if this is really a supplier problem or if our project manager is not giving our suppliers enough notice of when materials will be needed.

H

CHAPTER 9 | **HUMAN RESOURCES**

9

HUMAN RESOURCES MANAGEMENT

The human resources management section of the PMP exam focuses heavily on organizational structures, roles and responsibilities of the project manager, team building, and conflict resolution. It has questions from the *PMBOK® Guide* as well as several of the publications listed in the bibliography to this study guide (Chapter 15, Appendix D).

This knowledge area stresses the importance of the project manager's ability to manage and develop a team of individuals in a project setting.

Many of the definitions used in this knowledge area, although not widely used in many organizations, have been seen in project management literature for many years. It is important to memorize the definitions from this chapter.

You must also understand the various organizational structures, the experience and educational requirements of the project manager, types of power exercised by the project manager, and conflict management concepts.

Although both the administrative and behavioral aspects of human resources management are covered in the *PMBOK® Guide*, the exam seems to emphasize the behavioral aspects.

> **EXAM TIP**
> The human resources knowledge area does not have any monitoring or controlling processes.

The project manager has the responsibility to ensure that the project team members follow ethical behavior as part of developing and managing the team.

Things to Know

1. The four processes of human resources management:
    *   **Plan Human Resource Management**
    *   **Acquire Project Team**
    *   **Develop Project Team**
    *   **Manage Project Team**
2. How to develop a **responsibility assignment matrix**

3. The **human resource management plan**
4. **Project manager roles** and **responsibilities**
5. The **staffing management plan**
6. An approach to **acquiring the right team**
7. The project manager, **interpersonal skills**, and the **work environment**
8. The four **leadership styles**
9. Types of **power**
10. **Team building activities**
11. Five stages of **team development**
12. Characteristics of an **effective team**
13. The **team building process** and **barriers to team building**
14. **Sources of conflict** and **ways to manage conflict**
15. Six **motivational theories**
16. The differences between **team performance assessments** and **project performance appraisals**

Key Definitions

Authority: the right to make decisions necessary for the project or the right to expend resources.

Colocation: project team members are physically located close to one another in order to improve communication, working relations, and productivity. Also known as a "tight matrix" organization.

Leadership: the ability to get an individual or group to work toward achieving an organization's objectives while accomplishing personal and group objectives at the same time.

Organizational breakdown structure (OBS): different from a responsibility assignment matrix. The OBS is a type of organizational chart in which work package responsibility is related to the organizational unit responsible for performing that work. It may be viewed as a very detailed use of a RAM with work packages of the work breakdown structure (WBS) and organizational units as its two dimensions.

Power: the ability to influence people in order to achieve needed results.

Resource calendar: a calendar that documents the time periods in which project team members can work on a project.

Responsibility assignment matrix (RAM): a structure that relates project roles and responsibilities to the project scope definition.

Team building: the process of getting a diverse group of individuals to work together effectively. Its purpose is to keep team members focused on the project goals and objectives and to understand their roles in the big picture.

Virtual teams: groups of people with shared objectives who fulfill their roles with little or no time spent meeting face to face.

PLAN HUMAN RESOURCE MANAGEMENT PROCESS

This process involves the identification, assignment, and documentation of roles, responsibilities, and reporting relationships. Therefore, it is closely linked with the Plan Communications Management process, since the organizational structure influences communications requirements.

Although organizational structures are discussed in Chapter 2 on Project Management Overview and in the equivalent chapter in the *PMBOK® Guide*, organizational structure does influence organizational planning and communications planning. Additional exam questions on functional, matrix, and projectized organizational structures show up within this process because the organizational structure of the performing organization could be a constraint to the project team's options.

Enterprise environmental factors play a key role in the latitude a project manager has in acquiring and maintaining a competent project team.

Responsibility Assignment Matrix (RAM)

Organizational charts are useful to project managers in many ways. The *PMBOK® Guide* outlines three types of charts: **hierarchical**, **matrix-based**, and **text-oriented**. The RAM is a two-dimensional matrix-based chart relating group or individual roles and responsibilities to project work. The RAM may be at a high level, showing group or unit responsibility, or it may be at an individual or schedule activity level. A RAM can establish functional responsibilities, contracting strategies, and manageable work packages for control and reporting.

Figure 9-1 shows a lower-level RAM in which roles and responsibilities for various phases are assigned to particular individuals using the following assignments:
- Identification of those who perform the work
  P = Participant
- Identification of those who must approve
  A = Approval
- Specialized responsibility
  R = Responsible
- Identification of those who may be consulted
  I = Input Required
- Identification of those who must be notified
  RR = Review Required

Figure 9-1
Sample RAM

Person/Phase	Mary	Ivan	Tim	Erica	Allan	Jannette	Jose
Business Requirements	RR	R	A	P	P	—	—
Functional Requirements	RR	P	A	P	R	P	—
Design	RR	P	I	I	R	P	P
Development	I	—	—	R	R	—	I
Testing	R	A	I	P	RR	RR	P

P = Participant   A = Approval   R = Responsible   I = Input Required   RR = Review Required

RACI is a commonly used RAM format that differs from Figure 9-1 in that its criteria for responsibility are responsible (R), accountable (A), consult (C), and inform (I). The *PMBOK® Guide* advocates using the **organizational process assets** already in your organization.

Case Study Exercise

Exercise 9-1: Using the template below, create a RAM for the Lawrence Garage Project.

Responsibility Assignment Matrix							

P = Participant
R = Responsible
A = Approval
I = Input Required

Human Resource Management Plan

The human resource management plan provides information on project **roles**, **responsibilities**, **authority**, **competency**, project organization, staffing, and how human resources will be managed, controlled, and released. The **staffing management plan** is a subset of the human resource management plan which focuses on staff acquisition and the timing of staff needs.

Role and responsibility assignments (often in the form of a RAM), the staffing management plan (often in the form of a **resource histogram**), and the project organizational chart (sometimes including an organizational breakdown structure) are included in the human resource plan, the only output of this process.

EXAM TIP

A typical human resource management plan will include a:
- RAM
- Resource histogram
- Project OBS
- Resource acquisition approach

A **project manager's roles**: in the course of managing a project, the project manager will hold many roles. Some of the most important roles (from PMI's *Principles of Project Management*, pages 178 to 180) include being a/an:
- Integrator, who produces the product with available resources within time, cost, and performance constraints
- Communicator, who interfaces with customers, stakeholders, upper management, project participants, and functional managers
- Team leader, who is a team builder
- Decision maker, who makes or ratifies all required project decisions
- Climate creator or builder, who resolves conflicts

A **project manager's responsibilities**: the project manager's tasks and responsibilities include:
- Planning, scheduling, and estimating
- Analyzing costs and trends
- Reporting progress and analyzing performance
- Maintaining client-vendor relationships
- Managing logistics
- Controlling costs
- Handling organizational and resource issues
- Handling procedural, contractual, material, and administrative issues

One of the most important outputs of the Plan Human Resource Management process should be the documentation of the project manager's **authority**. It should be published to delineate his or her role in regard to:
- Point of contact for project communication
- Resolving conflicts
- Influence to cut across functional and organizational levels
- Major management and technical decision making
- Collaborating in obtaining resources
- Control over allocation and expenditure of funds
- Selection of subcontractors

Staffing Management Plan

A staffing management plan, together with the project organization chart and the roles and responsibilities needed to complete the project, is an integral part of the Plan Human Resource Management process. The staffing management plan:
- Describes how the project's human resource needs will be met and the timing of these resources
- Includes how and where staff will be acquired
- Describes the timetable and staff hours required of team members
- Defines the **release criteria** for team members
- Identifies team member training needs
- Provides clear criteria for recognition and rewards
- Could include strategies for compliance with various regulations, contracts, or policies
- Contains safety policies and procedures

EXAM TIP

Competency is the skill and capacity required to complete assigned activities within the project constraints. It should be considered for all project team members, including the project manager.

ACQUIRE PROJECT TEAM PROCESS

This process involves obtaining the people to work on the project. In many cases it is the project manager who must obtain the human resources for a project. Therefore, a key technique of this process is **negotiation**. Human resources can be acquired from functional groups, other project teams, or new hires. In addition, resources may be obtained from outside the performing organization, as discussed in Chapter 12 on project procurement management.

Acquiring the Right Team

There are five basic requirements for conducting a successful project. In order of importance, they are:
- Choosing the right people
- Finding people with a positive attitude
- Obtaining people with the appropriate skills
- Setting up the right organization
- Using the right methods

EXAM TIP

When acquiring staff, you might want to consider the following:
- Availability
- Cost
- Experience
- Ability
- Knowledge
- Skills
- Attitude
- International factors like time zones and language

In order to acquire the people with the appropriate skills, project managers find it is becoming commonplace to work with **virtual teams**. The lower costs and improved technology of electronic communications such as email, electronic meetings, and video conferencing make virtual teams feasible. However, **team building** becomes more challenging when team members spend little or no time meeting face-to-face. Frequent and regular communication becomes critically important for a virtual team to work. Ground rules for communication must be set with clear objectives, shared goals, protocols for conflict and issues resolution, involvement of the team in decision making, and recognition of individual and team successes.

The ground rules for the method of team member interactions should be specified in the communications plan for the project. Therefore, communications planning is closely linked to the acquisition of team members for the project.

Acquiring the project team is truly an iterative process. As **progressive elaboration** is applied to the scope and schedule, new requirements for resources may be uncovered. It is important for the project manager to continually assess the human resource needs of a project.

DEVELOP PROJECT TEAM PROCESS

This process involves all aspects of improving the interactions and interpersonal relationships of the project team members to enhance project performance. As feelings of trust, cohesiveness, and teamwork increase among team members, the project team improves its ability to achieve project objectives. **Training** and development of team members, setting **ground rules**, implementing **team building activities**, and **colocating** team members (whenever possible) are important techniques of this process.

In this process, the project manager must use various interpersonal skills such as leadership, communication, and motivation, to ensure successful team building.

EXAM TIP
Read *Appendix X-3* within the *PMBOK® Guide* on interpersonal skills.

Personnel assessment tools are also useful to the project manager and the project team in gaining a better understanding of the strengths and weaknesses within a project team and increasing team productivity.

A concerted team building effort should be initiated at the start of every project. However, team building is a continuous process due to the arrival and departure of project participants and the alteration of **roles** and **responsibilities** over the **project life cycle**.

Project Manager's Interpersonal Skills and Work Environment

Many equate leadership with those who hold a senior-level title within the organization. **Leadership** requires creating a vision for your team and planning for and executing that plan to reach those goals. Included in the responsibility of leaders is providing support to ensure that the team has every opportunity to succeed.

Project mangers ARE leaders and must use a variety of skills, including influence, to break down barriers and gain support, even without a senior-level title.

Interpersonal skills: in order to manage projects successfully, project managers require many skills. Some of the key skills for a project manager to know are:
- Communications skills, which include the ability to adapt to the audience being communicated to
- Leadership skills, which include the ability to see the big picture and use creativity and vision to help the team achieve its goals
- Decision-making skills, which include the ability to assess the current environment and develop ways for the team to reach consensus and make decisions
- Influencing skills, which include the ability to influence both the team and external parties when issues and conflicts arise to solve problems
- Political and cultural awareness includes the ability to understand and plan accordingly for various organizational political environments as well as the international cultural differences present in current projects
- Team building skills, which include the ability to bring a group of individuals together to achieve a common goal, identify when skills of the team are lacking, and develop plans to fill those gaps

EXAM TIP

Read the section on key interpersonal skills at the end of each chapter of this study guide to see how these skills can apply to a project.

9

Work environment: time and stress are two factors that can enhance or diminish performance. The project manager faces greater time challenges than most functional or operational managers. A delay of one or more critical tasks could delay the entire project; therefore, the project manager must be able to **influence** groups and individuals to get things done. The difficulty associated with assigning priorities for work can place managers under continuous stress. Stress can be used as a driving factor in enhancing productivity, but long-term stress often leads to poor performance and ill health.

The common characteristics of a project manager's work environment are:

- Extensive contact with people; the project manager is an integrator, which requires intense interaction with people
- Fast pace; the project manager is under high pressure to deliver within the defined schedule and cost requirements, which often leads to working longer hours
- Risk identification and vigilance; the project manager must constantly look to the future for upcoming factors or triggers that could positively or negatively impact project deliverables

Leadership Styles

There are four basic leadership styles that are typically found in organizations today (from Verma, *Managing the Project Team*, pages 146 to 147). These are:

- **Autocratic and directing,** in which decisions are made solely by the project manager with little input from the team
- **Consultative autocratic and persuading**, in which decisions are still made solely by the project manager with large amounts of input solicited from the team
- **Consensus and participating**, in which the team makes decisions after open discussion and information gathering

• **Shareholder and delegating** (otherwise known as laissez-faire or hands off), often considered a poor leadership style in which the team has ultimate authority for final decisions but little or no information exchange takes place

A combination of styles may be necessary or appropriate in a given situation. A trend in organizations today is to use a collaborative approach in which team members take on leadership roles as needed.

Authority and **power** are related yet different. Project managers cannot be effective with authority alone. A certain level of power, or influence, over others is needed.

Types of Power

EXAM TIP
Reward and expert power are recommended over coercive power.

There are two types of power a project manager can use: legitimate (positional) power or personal power. These types of power are further broken down as follows (from Verma, *Human Resource Skills*, page 233):
  • **Formal**: a legitimate form of power based on a person's position in an organization
  • **Reward**: a legitimate form of power based on positive consequences or outcomes a person can offer; it can also result from personal influence
  • **Coercive (penalty)**: a legitimate form of power based on negative consequences or outcomes a person can inflict; it can also result from personal influence
  • **Referent**: a personal form of power based on a person's charisma or example as a role model (an earned power)
  • **Expert**: a personal form of power based on the person's technical knowledge, skill, or expertise in some subject (an earned power)

The project manager may experience **power** and **authority** problems for a variety of reasons, including:
  • Power and authority not being perceived in the same way by everyone
  • Poor documentation or lack of formal authority for the project manager

- Dual or multiple accountability of team members
- A culture that encourages individualism instead of teamwork
- Vertical or stove-pipe loyalties instead of cross-organizational structures
- The inability to influence or administer rewards and punishments

Team Building Activities

Team building activities are a key technique in the Develop Project Team process. Team building is one of the many challenges faced by project managers. Teamwork improves overall performance, boosts team members' satisfaction, and reduces stress. It must be practiced consistently and frequently throughout the life cycle of a project. Some ground rules for effective team building are:

- Start team building activities early
- Make sure all contributors to the project, whether full or part time, are included as part of the team
- Plan team building activities by phase or other major project or team change; team building must be reinitiated and repeated after the occurrence of any serious risk events or change in project direction
- Recruit the best possible people
- Obtain team agreement on all major actions and decisions
- Communicate as frequently and openly as possible
- Recognize that team politics exist, but do not take part in or encourage them
- Be a role model
- Encourage and mentor team members
- Evaluate team effectiveness often
- Use proven and effective team building techniques
- Quickly move ineffective team members to positions that match their skills

The **five stages of team development** are:
- **Forming**: team members come together and learn about their roles and responsibilities
- **Storming**: team members express their ideas and may be in disagreement about their approach to a project
- **Norming**: team members settle in and accept their roles; they agree on the approach to the project and how to work together
- **Performing**: team members accomplish their tasks and resolve conflict and issues effectively
- **Adjourning**: team members complete their work and leave the project

Characteristics of an effective team are:
- Team members are interdependent
- Team members have reasons for working together
- Team members are committed to working together
- The team as a whole is accountable
- Level of competition and conflict are manageable

In the **team building process**, the project manager carries out the process of getting a group of individuals to work together effectively by utilizing the following methods:
- Planning for team building by clearly defining project roles and making sure project goals and members' personal goals coincide
- Negotiating for team members by obtaining the most promising members for technical knowledge and potential to be effective collaborators
- Organizing the team by matching assignments to skills, and creating and circulating the project RAM
- Holding a kickoff meeting in which team members meet, technical and procedural guidelines are set, and work relationships and communications plans are established
- Obtaining team member commitments
- Building open and frequent communications links
- Incorporating team building activities into project activities

---

EXAM TIP

Allowing the project manager to select key team members helps accelerate the time it takes the team to reach the "performing" stage of team development.

9

Barriers to team building include:
- Differing outlooks, priorities, and interests
- Unclear project objectives or outcomes
- Dynamic project environments
- Lack of team definition and structure
- Role conflicts
- Poor credibility of a project leader
- Lack of team members' commitment
- Communication problems
- Lack of senior management's support

EXAM TIP
Whenever possible, plan team building activities early on in the project life cycle and continue such activities throughout the project to maintain morale and momentum.

MANAGE PROJECT TEAM PROCESS

This process involves **observing** and **tracking** team behaviors and performance, providing feedback, managing issues and conflict, and providing input to organizational performance appraisals. Progressing toward project deliverables, resolving conflicts and issues, and project performance appraising are the responsibility of the entire project team.

Team performance assessments from the Develop Project Team process, together with **work performance reports**, are some of the key inputs to this process.

To successfully manage project teams, project managers must use a variety of tools and techniques as the project warrants such as **observation** and **conversation**, **project performance appraisals**, **conflict management**, and **interpersonal skills**.

Sources of Conflict

Conflicts are unavoidable on projects due to projects' temporary nature. These conflicts can arise from:
- Projects being carried out in high-stress environments
- Roles and responsibilities being unclear or ambiguous
- The multiple-boss syndrome in which the priority of work becomes an issue
- Technologies being new or complex
- Teams being brought together for the first time

EXAM TIP
The process of managing conflict within the project team is initially the responsibility of the project team members

9

The seven main sources of conflict have been demonstrated to vary significantly depending on the phase of the project, but during a project's life cycle, the seven sources of conflict can be ranked as follows (from Verma, *Human Resource Skills*, page 102):

- Schedule issues
- Priority of work issues
- People resource issues
- Technical options and performance trade-off issues
- Administrative procedures
- Interpersonal relationship issues
- Cost and budget issues

Ways to Manage Conflict

The contemporary view of conflict management is that conflict can have a positive or negative impact on a project and on an organization. It can and should be managed. Whether it is beneficial or detrimental to the project depends on the source of the conflict and the way it is handled by the project manager. Conflict is also a natural result of change and is inevitable when people work together.

Should a disruptive conflict continue or escalate, the project manager must become involved and help find a satisfactory solution to the conflict. The project manager needs to apply various motivational techniques to overcome any such issues.

Conflict resolution is situational. While there are preferred methods, project managers may handicap themselves by using a single resolution method in all circumstances. The five ways to manage conflict (from Verma, *Human Resource Skills*, page 139) are problem solving, compromising, forcing, smoothing, and withdrawal.

In a **problem solving** or **confrontation** method of resolving conflict, the project manager directly addresses the disagreement and gets all parties to work together and want to solve the problem. The problem is defined, information is collected, alternatives are identified, and the most appropriate solution is selected. This method is considered win-win and is recommended for long-term resolution.

In using a **compromising** method, various issues are considered and a solution that brings some degree of satisfaction to the conflicting parties is agreed on. Both parties give up something that is important to them. This method is considered lose-lose and is likely to be temporary.

Project managers can also use **forcing** to manage conflict. In forcing, one person's viewpoint is exerted at the expense of another party. This method is considered win-lose and it can build antagonism and cause additional conflicts. It may be appropriate in low-value situations.

In **smoothing** a conflict, the opposing party's differences are deemphasized and commonalities are emphasized on the issue in question. This keeps the atmosphere friendly, but this method does not resolve the conflict; it only delays it. Smoothing could be used with one of the above three methods.

A project manager who uses **withdrawal** is retreating from the actual or potential issue or conflict situation. This method is appropriate only for situations in which a cooling off period is needed. Withdrawal does not resolve the conflict; it only delays it.

Motivational Theories

There are six motivational theories that are emphasized on the exam (from Verma, *Human Resource Skills*, pages 70 to 75). These six theories were developed by Maslow, Herzberg, McGregor, Ouchi, Vroom, and the team of Hersey and Blanchard.

Maslow's hierarchy of needs theory is a model of needs in the form of a pyramid with five levels. He suggests that within every person there resides a hierarchy of needs that duplicates the levels of the pyramid. The needs of the lower level of the pyramid must be satisfied before higher level needs can be addressed. Motivation springs from needs that are not met, as the person strives to fulfill these needs. The levels of needs in this pyramid are, in order of the lowest to highest:

- **Physiological**, the need for food, shelter, and items for survival
- **Safety**, the need to be safe from danger, threat, and deprivation
- **Social**, the need for association with humans, friendship, and acceptance
- **Self-esteem**, the need for self-respect, status, and respect from others
- **Self-actualization**, the need for self-fulfillment through the development of powers, skills, and one's own creativity

Herzberg's theory of motivation relates Maslow's theory of needs to the workplace. Herzberg suggested that motivation in one's work is the result of two factors:

- **Hygiene**, such as the attitude of a supervisor or working conditions; poor hygiene factors may destroy motivation but improving hygiene factors under normal circumstances is not likely to increase motivation (for example, a clean working environment does not motivate, but a dirty work environment will demotivate)

- **Motivators**, such as interesting work, opportunities for personal growth, achievement, and recognition; such positive motivators offer opportunities to achieve and experience self-actualization because they give workers a sense of personal growth and responsibility; depending on the organizational culture, group success may be more important than individual achievement.

McGregor's Theory X assumes that workers need to be constantly watched and told what to do. Managers who subscribe to Theory X believe that the average worker dislikes work and avoids work whenever possible. This worker is only motivated by money, position, and punishment. In addition, a worker avoids increased responsibility and seeks to be directed. Therefore, to induce adequate effort, the manager must threaten punishment and exercise careful supervision. The manager who practices Theory X normally exercises authoritarian-type control over workers and allows little participation by workers in decision making.

McGregor's Theory Y assumes the opposite set of characteristics about human nature. Managers who subscribe to Theory Y believe that workers are self-disciplined and will do a job themselves. The average worker wants to be active in a supportive work climate and finds the physical and mental effort on the job satisfying. The greatest results come from willing participation, which will tend to produce self-direction towards goals without coercion or control. The average worker seeks opportunities for personal improvement and self-respect. The manager practicing Theory Y normally advocates a participatory management-employee relationship.

9

Ouchi's Theory Z is built off of McGregor's Theory X and Theory Y. Ouchi postulated that quality does not lie with technology but rather lies in a special way of handling people. His theory is based on the Japanese cultural values of lifetime employment, slow promotions, nonspecialized career paths, and collective decision making. Theory Z postulates that high levels of trust, intimacy, confidence, and commitment to workers by management result in high levels of motivation and productivity by workers.

Vroom's expectancy theory postulates that people think about the effort they should put into a task before they do it; if workers believe their efforts are going to be successful and rewarded, they will tend to be highly motivated and productive.

In **Hersey and Blanchard's life cycle theory**, the leadership style must change with the maturity of individual employees; maturity is defined as the extent of job-related experience. In this theory, the situation drives the leadership style to be used to motivate each worker. Generally, the project manager's style should move from directing, to coaching, to supporting, then to delegating as the project moves through its life cycle

Project Performance Appraisals versus Team Performance Assessments

As part of the Manage Project Team process, the project manager should be involved in project performance appraisals. The focus here is on evaluating how individuals are performing within the context of the project, providing feedback, and identifying ways to enhance team members' performance through training.

Team performance assessments, a tool and technique of the Develop Project Team process, differ from project performance appraisals in that they are focused on the evaluation of the team's effectiveness in meeting the objectives of a project. By design, these are not individual assessments. The results of a team performance assessment may be recommendations for improvement for individuals, but these assessments are focused on the overall team's performance and on assessing how the team can improve its ability to perform as a group.

Motivating people to perform work to the best of their ability is a challenge faced by every project manager. Compensation in terms of money is not enough; other methods must be considered, some of which are:

- **Fringe benefits** in education, profit sharing, medical benefits, etc.
- **Perquisites** or **perks** such as a parking space, window office, or company car
- **Arbitration** or **dispute resolution** using a third party to resolve conflicts
- **Career planning** outlining possibilities for growth
- **Training** in new skills
- **Productivity incentives**
- **Team camaraderie**

EXAM TIP
Project managers who are involved in team performance assessments increase the reward and penalty power over team members.

9

SAMPLE PMP EXAM QUESTIONS ON HUMAN RESOURCES MANAGEMENT

1. The process which addresses how to obtain the human resources necessary for a project is:

   a) Manage Project Team
   b) Develop Project Team
   c) Plan Human Resource Management
   d) Acquire Project Team

2. As the project manager of a software development project, your lead technical architect and lead programmer typically do not see eye-to-eye on technical approaches. In the weekly status meeting, their discussions have escalated into conflict. One of the team members has questioned whether or not the meeting has been productive. You should:

   a) Discuss the concerns with the team as a group and come to a consensus as to how to address them
   b) Discuss the concerns with the technical architect and programmer either together or separately but away from the rest of the project team
   c) In the meeting, limit the amount of time allocated to the lead technical architect and lead programmer have to discuss the issue
   d) Don't include the architect in the next team meeting so that the programmer will have time to air the issues with the group

3. As the project manager of a new project, you understand the importance of team development. Because of this understanding, you plan a team building event:

   a) When the sponsor is able to attend, which will increase the team's visibility with the sponsor
   b) Early in the project and then again at regular intervals during the project
   c) At the beginning of the project to have all team members meet each other
   d) When human resource risk triggers occur

4. To ensure that each team member has a clear understanding of his or her roles and responsibilities, the project manager should:

   a) Ask the person doing the work for best and worst case estimates
   b) Ensure that each work package has an unambiguous owner
   c) Develop an organizational breakdown structure
   d) Develop a project organization chart

5. Assessments of the team's performance that show it is getting better can include:

   a) Improvements in individual skills or competencies
   b) Increased staff turnover rate
   c) Additional storming behaviors
   d) Missing schedule deadlines

6. The main benefit of the process of documenting project roles and responsibilities in a responsibility assignment matrix is that:

   a) It reduces the cost of resources assigned to the project
   b) It helps the project manager make sure that stakeholder expectations are met
   c) It clarifies and communicates who is contributing to and responsible for each project activity
   d) It can be used as an input to planning the procurement of resources

Notes:

9

7. There is a motivational theory that states the belief that under the right conditions, workers enjoy work, will seek responsibility, and perfer to direct their own work. This theory is called:

   a) McGregor's Theory X
   b) Maslow's Hierarchy of Needs
   c) MacGregor's Theory Y
   d) Herzberg's Motivation Theory

8. Influencing team behavior, resolving issues, and appraising the performance of individual team members is a key benefit of which process?

   a) Acquire Project Team
   b) Plan Human Resource Management
   c) Develop Project Team
   d) Manage Project Team

9. A subcontractor is two weeks late in his deliverables and asks the project manager to accept these late deliverables in exchange for a reduction in his fee. This is an example of the _____ technique of conflict resolution.

   a) Forcing
   b) Problem solving
   c) Compromising
   d) Withdrawal

Notes:

10. As the project manager of a team that is acting dysfunctionally, you put together a plan to increase face-to-face gatherings and team building activities. An additional option you may pursue to better individual team members' performance is:

    a) Asking the team members to actively disagree with other team members
    b) Creating a competitive envirnoment within the project team
    c) Establishing a team reward for the completion of major milestones
    d) Recognizing the top performer in the team each month

11. You are the project manager on a government energy contract. Government regulations stipulate that you must ensure that more than 30% of all subcontractors be from minority- or woman-owned businesses. As the project manager, this information should be included in the:

    a) Cost management plan
    b) Responsibility assignment matrix
    c) Staffing management plan
    d) Contract templates

12. As the project manager of a website development project with limited resources, you discover that several team members have no experience. You decide that the best way to develop your project team is to:

    a) Team up experienced and inexperienced team members
    b) Take the team to a kick-off luncheon
    c) Include team training in the project schedule and staffing management plan
    d) Hold regular team meetings to review project progress

Notes:

13. One of your team members, John, has misinterpreted work package instructions and has not produced the appropriate deliverables on time or without significant rework being required. This issue has impacted the overall schedule of the project negatively. For the next work package to be assigned to John, you should:

   a) Talk to John and explain the firm deadlines for the project and the need to keep cost under control
   b) Talk to John and have him communicate his understanding of the work package deliverable
   c) Inform your sponsor that you are removing John from the team
   d) Ask John for revised estimates on when his work will be completed

14. What impact does a matrix organization have on project team management and development?

   a) Team development is simplified
   b) Team development becomes more complex
   c) There is no impact
   d) Team development does not take place in matrix organizations

15. You are the project manager for a business process improvement project. Ava and William are experts in the critical business process being redesigned, but they work at a different facility from the rest of the project team. This situation poses additional challenges to the project manager and requires the use of tools and techniques for:

   a) Virtual teams
   b) Time management
   c) Quality management
   d) Integration

Notes:

9

ANSWERS AND REFERENCES FOR SAMPLE PMP EXAM QUESTIONS ON HUMAN RESOURCES MANAGEMENT

Section numbers refer to the *PMBOK® Guide.*

1. **D** **Section 9.2 – Executing**
   The Acquire Project Team process is an executing process.

2. **B** **Section 9.4.2.3 – Executing**
   This is an example of managing conflict. A) and C) are not the best answers as they may put the architect and programmer in a defensive position; D) not including team members does not resolve conflict or instill confidence.

3. **B** **Section 9.3.2.3 – Executing**
   Team building should be done throughout the project life cycle. It is especially important early on to build rapport between team members and to gain maximum cohesiveness.

4. **B** **Section 9.1.2 – Planning**
   A) asking for best or worst case estimates does not ensure understanding; C) an organization breakdown schedule relates work packages to the responsible organization, not to the team member; D) a project's organization chart reflects reporting relationships, not responsibility.

5. **A** **Section 9.3.3.1 – Executing**
   Assessing team performance can be informal or formal. Effective team development activities will increase team performance.

6. **C** **Section 9.1 – Planning**
   A) the risk of misunderstanding responsibility may be reduced, and the costs are not necessarily reduced; B) indirectly, a RAM may help assure that stakeholder expectations are met by having the right people in the job, but this is not the best answer; D) usually a RAM will be prepared after the make-or-buy decision is made.

7. **C** **Section 9.3 – Executing**
Motivation plays a big part in individual contributions and team development.

8. **D** **Section 9.4 – Executing**
The *PMBOK® Guide* identifies the key benefit for each process in a Knowledge Area. This helps you understand what is going on in that process.

9. **C** **Section 9.4.2.3 – Executing**
Both parties give up something; therefore, it is a compromise. Often, this method of resolving conflict is called lose-lose.

10. **C** **Section 9.1.3.1 – Planning**
A) when a team is dysfunctional, it may be better to have people find ways to agree, not disagree; B) a competitive environment may increase the teams' dysfunciton; D) recognizing the top performer focuses on individuals, not on the the team's performance.

11. **C** **Section 9.1.3.1 – Planning**
The staff management plan must take into consideration compliance with applicable government regulations and human resource policies.

12. **C** **Section 9.3.2.2 – Executing**
Because there are limited resources, it may be difficult to team people up, so C) is the best answer to address the issue presented.

13. **B** **Section 9.4.2.2 – Executing**
The best answer is to ensure that the next task John works on is clearly communicated and John understands fully his responsibilities and deliverables. Asking for revised time estimates will not solve the problem of poor work.

14. B Section 9.3 – Executing
Team members in a matrix organization are accountable to both the functional manager and the project manager, which can cause conflicting loyalties.

15. A Section 9.2.2.4 – Executing
Remote team members can impact how a project manager manages time, quality, and integration, and it REQUIRES that a project manager leverage tools and techniques for virtual teams.

9

CASE STUDY SUGGESTED SOLUTION

Exercise 9-1:
RAM for the Lawrence Garage Project

Responsibility Assignment Matrix	Project Management	Site Work	Foundation/Slab	Framing	Dry-in	Exterior	Rough Utilities	Interior	Grounds	Acceptance
Concrete crew			P						P	
Finish carpentry crew								P		
Lath & plaster contractor						P				
General contractor	A	R	R	R	R	R	R	R	R	R
Architect	A	I	I	I	I	I	I	I	I	I
Owner	A	I	I	I	I	I	I	I	I	A
Inspector		A	A	A	A	A	A	A	A	A
Electric company		P					P			
Site excavation crew		P							P	
Electric contractor		P					P	P		
Plumbing contractor							P	P		
Framing crew				P	P	P	P			
Roofing contractor					P					
Heating and AC contractor							P	P		
Insulation contractor								P		
Drywall crew								P		
Painting crew								P		

P = Participant
R = Responsible
A = Approval
I = Input Required

9

COMMUNICATIONS

CHAPTER 10 | **COMMUNICATIONS**

10

10

COMMUNICATIONS MANAGEMENT

One of the primary and most important roles of the project manager is communication—communication of project objectives, management strategies, and the project plan. A project manager's responsibility is to facilitate understanding, thereby enhancing the team's effectiveness.

Inadequate communication, one-way communication, incomplete messages, and unclear messages are common problems in many projects. Communication skills fall under both general management and project management skills and are necessary for the effective exchange of information. The project manager has a responsibility to:
- Know what kind of message to send to what audience
- Know what format and method to use for each message
- Determine the timing of the message

> **EXAM TIP**
> 90% of the project manager's time is spent communicating (from Kerzner's *Project Management: A Systems Approach to Planning, Scheduling, and Controlling*, page 232).

Many of the questions on the Communications Management processes in the PMP Exam are taken from the *PMBOK® Guide*. There will be questions on specific terms and concepts, but there will also be many general questions that require you to choose the best answer. Common sense and your own experience will play a large role in your ability to answer questions on this topic.

You will most likely have questions related to formal, informal, push, pull, and interactive communication, as well as communication models and technology. There is also an emphasis on the types of enterprise environmental factors and organizational process assets that are inputs or outputs to communications management processes.

Things to Know

1. The five processes of communications management:
   - **Plan Communications Management**
   - **Manage Communications**
   - **Control Communications**
2. **Communication requirements analysis**
3. The **communication channels** formula
4. Use of **communication technology**
5. The **communication model**
6. **Barriers to communication**, such as **language** and **culture**
7. **Communication methods**
8. What the **communications management plan** is
9. **Information management systems**
10. Purpose of **performance reporting**
11. Key **interpersonal skills** for success

Key Definitions

Acknowledge: indicates receipt of a message by a receiver, but does not indicate that the receiver understood or agreed.

Active listening: the receiver confirms listening by nodding, eye contact, and asking questions for clarification.

Decode: the term for the receiver translating a message into an idea or meaning.

Effective listening: the receiver attentively watches the sender to observe physical gestures and facial expressions. In addition, the receiver contemplates responses, asks pertinent questions, repeats or summarizes what the sender has sent, and provides feedback.

Encode: the term for the sender translating an idea or meaning into a language for sending.

Feedback: affirming understanding and providing information.

EXAM TIP

Active listening is a valuable skill to learn for all project managers. It ensures that the message being communicated is understood.

10

Noise: anything that compromises the original meaning of a message.

Nonverbal communication: about 55% of all communication, based on what is commonly called body language.

Paralingual communication: optional vocal effects, the tone of voice that may help communicate meaning.

Transmit message: the term for using a communication method to deliver a message.

PLAN COMMUNICATIONS MANAGEMENT PROCESS

During the Plan Communications Management process, the stakeholders' communication requirements are determined. These informational needs are documented in the communications management plan and take into consideration requirements such as:
- Who needs what information?
- When will they need it?
- How will it be given to them?
- Who will give it to them?
- How will information be stored and retrieved?

Communication Requirements Analysis

This tool is intended to focus on project stakeholders' needs for information and should include the type and value of information and how it will be presented. A key input to this analysis is the **stakeholder register**.

The project manager must take into consideration and analyze all of the information provided to stakeholders across the organization, as well as information external to the organization.

Understanding the logistics of the project team is important. If teams are virtual, it is even more critical to ensure that communication between team members is covered within this analysis.

10

The Communication Channels Formula

There are many communication channels utilized by a project manager. These include:
- Upward communication to management
- Lateral communication to peers, other functional groups, and customers
- Downward communication to subordinates

Communication channels are the number of one-to-one communications that exist for the team. The more channels, the more complex the communications analysis. Figure 10-1 below shows communications channels among five people.

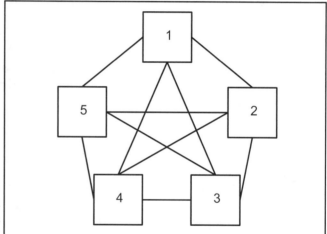

Figure 10-1
Communication
Channels

There is a simple formula to determine the number of communication channels that exist on a project:

$(n[n - 1]) \div 2$ where n indicates the number of people

For example, if 5 people work on a project, n = 5, communication channels = $5(4) \div 2 = 10$.

With 7 people, n = 7, communication channels = $7(6) \div 2 = 21$. The number of channels has more than doubled with just two additional team members.

Case Study Exercise

Exercise 10-1: How many communication channels are there on a project with 50 people?

Communication Technology

In analyzing communications, you must address the types of technologies available to facilitate communications and make their transfer more effective. Some considerations include the **urgency** of information receipt, the types of technology available, the **sensitivity** or **confidentiality** of the information, the amount of training required, the length of the project, and the number of **virtual team** members and stakeholders.

The Communication Model

A basic communication model consists of the following components: a sender, a receiver, a medium through which messages are sent and received, noise, and feedback. Figure 10-2 below shows the communication model.

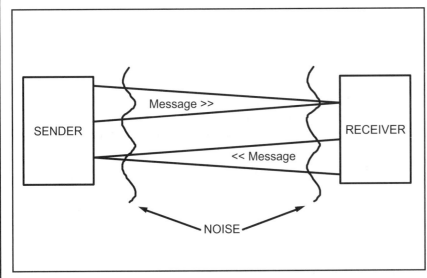

Figure 10-2
Communication
Model

The **sender** encodes a message, chooses the medium in which to send it, and attaches symbols, gestures, or expressions to confirm that the message is understood. As the message passes through the medium, it encounters "**noise**" that interferes with transmission and meaning. Noise can be many things—from physical noise in the background during a conversation to having something on your mind that is distracting you from paying full attention to the speaker. The **receiver** decodes the message based on the receiver's background, experience, language, and culture. The receiver sends a feedback message back through the medium and noise to the sender that may be an acknowledgement of receipt or a response indicating understanding.

Barriers to Communication

In addition to the large number of communication links required as resources increase, other barriers also exist to deter effective communication. Some of these barriers are:

- Ineffective listening
- Improper encoding of messages
- Improper decoding of messages
- Naysayers
- Hostility
- Language
- Culture

Language and **culture** are especially common sources of problems in communication. Communicating across cultures has many challenges for organizations as well as team members.

- High-context language messages require the reader or listener to know the situation—the context—that the message is discussing
- Low-context language messages do not require this knowledge; they contain all the information needed to understand them in the message itself; low-context communication is the more explicit and straightforward communication pattern

Figure 10-3 shows how some languages fall on a scale of low- to high-context languages.

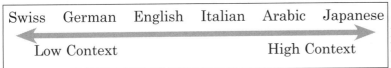

| Swiss | German | English | Italian | Arabic | Japanese |

Low Context   High Context

Figure 10-3
Low- to High-
Context Languages

Factors complicating effective communication across cultures include:
- Slang
- Euphemisms
- Proverbs, or wise sayings
- Humor
- Nonverbal communication
- Personal space
- Personal contact
- Individualism versus group cohesion

In addition, the project manager has a responsibility to ensure that no illegal, offensive, inaccurate, or incomplete communication is conducted.

Communication Methods

Communication methods may be broadly classified into three categories:
- **Interactive** methods include meetings, phone calls, and videoconferencing
  - **Formal verbal communication** such as presentations and speeches should be used when persuading people to accept ideas and products
  - **Informal verbal communication** such as meetings, conversations, humor, and inquiries are used for small informal groups, team building, and day-to-day communication
- **Push** methods include letters, memos, reports, emails, faxes, etc. that are sent to stakeholders
  - **Formal written communication** should be used for key documents such as project plans, the project charter, communicating over long distances, complex problems, legal documents, and long or technical situations for a wide and varied audience

> **EXAM TIP**
> Email, texting, and tweeting are informal communication methods according to the *PMBOK® Guide*.

10

- • **Informal written communication** should be used for status updates, information updates, and day-to-day communication
- **Pull** methods within an organization include intranet sites and knowledge repositories and are typically used when there are large volumes of information to share or a large audience to reach

Communications Management Plan

The communications management plan is an output of the Plan Communications Management process. It should be created by the project manager and becomes part of the **project management plan**. The communications management plan must also include the:

- • Stakeholder communication requirements
- • Information required
- • Method used to convey information
- • Reporting responsibilities
- • Distribution schedule
- • Performance reporting process

The use of a **communications matrix** can help identify and organize this information. Note that the **organizational structure** (functional, matrix, or projectized) will influence the information and distribution channels of a project. There are many formats for communications matrices. No matter what the format, the objective is to identify who is being communicated to, when communication is needed, how communications will be distributed, and who is responsible for their delivery.

Case Study Exercise

Exercise 10-2: Use the table on the following page to develop a sample communications matrix for the Lawrence Garage Project. Identify the different kinds of communication, the frequency of communication, who is responsible for creating each piece of communication, and who receives each communication.

Communications Matrix							

R = Report progress
P = Prepare report
A = Attend meeting
RR = Review report

MANAGE COMMUNICATIONS PROCESS

The Manage Communications process deals with the flow of information, especially work performance reports among project stakeholders. Project managers must consider how the sender-receiver model and choice of media can be used to make sure work performance reports and other project documents are shared with the appropriate stakeholders.

Communicating effectively to ensure that information is understood and that stakeholders give feedback relies on various techniques that include:
- Writing style
- Meeting management techniques
- Presentation techniques
- Facilitation techniques
- Listening techniques

Work Performance Reports

Work performance information that has been collected and pulled together as work performance reports is a key input to the Manage Communications process.

Types of performance reports include:
- Status reports
- Trend and forecasting reports
- Approved change requests
- Process updates
- Risk monitoring and control outcomes

Information Management Systems

Project managers have primary responsibility for the wide array of communication that takes place on a project. Their responsibility includes understanding the use and management of physical media (such as letters and reports), electronic communications (such as email and voice mail), and electronic tools (such as collaborative project management software), no matter where each resides.

EXAM TIP

Performance reports must be aligned with the stakeholder register.

10

Performance Reporting

Performance reporting is the act of collecting and distributing **work performance information**. Performance reporting should be appropriate for the stakeholders who receive the information. Such reporting can be used to determine progress, work completed, status, issues, action items, and forecasts. Performance reporting using earned value management, as discussed in Chapter 7, can be more effective when planned scope completed is compared to actual scope completed.

CONTROL COMMUNICATIONS PROCESS

The Control Communications process focuses on making sure stakeholders have the information they need at the right time and in the right format by monitoring and controlling how communications flow. The goal is to optimize the flow of information. What project managers learn during this process may show that there is a need for replanning communication strategies or changing how communication is managed.

> **EXAM TIP**
> The project manager should initiate discussions with stakeholders to ensure that all parties are aligned regarding what is to be delivered on the project.

Work performance data are an input to this process and work performance information is an output.

Earned value techniques are very helpful in providing answers to common time and cost performance questions such as:
- Where are we in the project schedule?
- What is the percentage of completion?
- What is the estimated time to complete the project?
- Are we ahead or behind schedule?
- What are actual expenditures to date?
- What are the committed expenditures?
- What are the estimated remaining costs?
- Are we under budget or over budget?

Responding to the answers to these types of questions is what the Control Communications process is all about. For instance, if a project is behind schedule, the project manager needs to know whom to let know about the delay and whom to contact to get the project back on track.

10

KEY INTERPERSONAL SKILLS FOR SUCCESS

The interpersonal skill highlighted in this chapter is:

Communication

Communications management focuses on managing information from its creation, flow, and ultimate archiving or disposition. There is an explicit acknowledgement that a project manager's personal skill at communicating is vital to a project's success. The project manager must work to understand others and their personal communication styles, cultural norms that affect communications, and their own relationships with others. Active listening techniques help a project manager build other interpersonal skills such as negotiating and influencing.

One of the project manager's primary responsibilities on a project is to ensure information is being used appropriately. The project manager must make sure that all of the following are documented, shared, and reviewed by all appropriate stakeholders:
  • Project vision
  • Schedules
  • Stakeholder expectations
  • Acceptance criteria
  • Project status

A key concept that is highlighted in the June 2015 *PMI PMP Exam Content Outline* is the fact that the project manager is responsible for executing on the communication plan to ensure all stakeholders thoroughly understand what the project is expected to deliver. This is accomplished by reviewing and gaining approval on the project charter and regularly assessing if the communication plan is adequate to ensure the project can be delivered successfully.

10

SAMPLE PMP EXAM QUESTIONS ON COMMUNICATIONS MANAGEMENT

1. The number of communication channels for a project with 15 stakeholders is _____.

   a) 45
   b) 105
   c) 150
   d) 90

2. You are a project manager on a hotel construction project. The CEO and sponsor of the project retired from the business and a new CEO has been named. The new CEO wants the team to provide a monthly formal presentation to the senior management team. You will need to:

   a) Throw out all prior status reports and create this new one from scratch
   b) Add the monthly deliverable to the project schedule and budget
   c) Hire an administrative assistant since you're not proficient with PowerPoint
   d) Bring the team together and assign team members to each of the agenda items for this review

3. Using a web interface to update the project schedule is an example of a technique in which process?

   a) Manage Communications
   b) Distribute Information
   c) Control Cost
   d) Plan Communications Management

Notes:

4. You are assigned as a new project manager of a project that is 40% complete. You are assigned due to the departure of the prior project manager. You review the current issues logs and meeting minutes. You talk to the project sponsor and other key stakeholders and determine that there are two primary reasons for the problems with the project. You should:

   a) Update the WBS
   b) Document the lessons learned
   c) Add schedule activities to the project schedule
   d) Add the issues to the risk register

5. An email is an example of which form of communication?

   a) Pull communication
   b) Interactive communication
   c) Push communication
   d) Formal communication

6. A key benefit of the Control Communications process is that:

   a) It ensures that information is flowing to all project stakeholders
   b) It identifies and documents approaches to communicate
   c) It ensures that the project manager is following the plan
   d) It allows for an efficient flow of communications

7. As the project manager on a software development project, you hold status review meetings regularly. Status review meetings are an example of a tool within the _____ process.

   a) Plan Stakeholder Management
   b) Manage Project Team
   c) Plan Communications Management
   d) Control Communications

Notes:

10

8. Which of the following enterprise environmental factors can influence the Manage Communications process?

   a) Choice of media and writing style
   b) Meeting management and facilitation techniques
   c) Organizational culture and government standards
   d) Templates and lessons learned

9. Work performance information that is organized and summarized includes:

   a) Information management systems to store and distribute information
   b) Meetings to determine ways to respond to stakeholder requests
   c) Expert judgment from customers and end users
   d) Status and progress information on a project

10. You are the project manager on a project in which 60% of all team members are offsite contractors and not employees, and the project is two months from completion. Your company has just implemented a new technology for internal communications. Should this technology be something that could be considered for your project team's use?

    a) Yes, because the system is company-wide
    b) No, because the there is already a communication system in place
    c) Yes, because of the virtual nature of the team
    d) No, since it would require training and access for all non-employees

11. Which of the following organizational process assets might influence the Control Communications process:

    a) Meeting management and facilitation techniques
    b) Organizational culture and government standards
    c) Report templates and record retention policies
    d) Information management systems and listening techniques

Notes:

12. When executing projects, communication typically focuses on reports related to:

a) Performance and deliverables status
b) Requirements documentation
c) Quality planning
d) Stakeholder identification

13. A change request that outlines steps recommended to bring expected future project performance in line with the project management plan is:

a) A new cost estimate
b) A recommended corrective action
c) Work performance information
d) A management reserve

14. A key input to the Control Communications process is:

a) Communication model
b) Stakeholder communication requirements
c) Staff management plan
d) Resolved issues

15. Communications should be monitored and controlled throughout the project life cycle in order to:

a) Increase the efficiency of stakeholder engagement activities
b) Have efficient communication among stakeholders
c) Meet stakeholder's information needs
d) Plan for communication that meets stakeholders' needs

Notes:

10

ANSWERS AND REFERENCES FOR SAMPLE PMP EXAM QUESTIONS ON COMMUNICATIONS MANAGEMENT

Section numbers refer to the *PMBOK® Guide*.

1. **B   Section 10.1.2.1 – Planning**
   The number of communication channels = $n(n-1) \div 2$ where n is the number of stakeholders.

2. **B   Section 10.3 – Monitoring and Controlling**
   A) and C) are not necessarily the case in all projects; D) is not a bad answer, but the question doesn't necessarily define the size and agenda items of the presentation; B) is the most correct answer because the development of the presentation is added work that will need to be performed on the project to support the stakeholder's needs.

3. **A   Chapter 10.2.2.3 – Executing**
   Information management systems are tools for the Manage Communications process that include hard copies of documents, electronic communications, conferencing, and project management tools; B) "distribute information" is no longer a process name; C) this question is about schedule, not cost; D) communication technology to be used during the project is defined during the Plan Communications Management process.

4. **B   Section 10.2.3.4 – Executing**
   It is important to document lessons learned as they occur. Since the issues have already occurred, they are not considered risks.

5. **C   Section 10.1.2.4 – Planning**
   Understand the differences in types of communication and when each is used.

6. A **Section 10.3 – Monitoring and Controlling**
B) is part of the Plan Communications Management process; C) there is no process that specifically addresses whether the project manager is following the plan; D) is part of the Manage Communications process.

7. D **Section 10.3.2.3 – Monitoring and Controlling**
Status review meetings are a communication method used to exchange and analyze information about a project.

8. C **Chapter 10.2.1.3 – Executing**
A) and B) are basic communications skills; D) are organizational process assets.

9. D **Chapter 10.3.3.1 – Monitoring and Controlling**
A), B), and C) are all tools and techniques of the Control Communications process.

10. D **Chapter 10.3.2.2 – Monitoring and Controlling**
Considering the project staffing and their location and access to communications systems is important. Since this project is within two months of completion, the amount of training and connectivity concerns would most likely outweigh the benefits of the new system.

11. C **Section 10.3.1.5 – Monitoring and Controlling**
A) are basic communications skills; B) are environmental enterprise factors; D) information management systems are environmental enterprise factors, and listening techniques are basic communication skills.

12. A **Section 10.2.3.1 – Executing**
B) and C) are planning processes; D) is an initiating process.

10

13. B Section 10.3.3.2 – Monitoring and Controlling
A) a new cost estimate might be required, but
B) is a better answer; C) work performance
information summarizes performance data that
has been collected; D) management reserves are
for risk events that are unknown unknowns.

14. B Section 10.3.1.1 – Monitoring and Controlling
A) is a tool for the Plan Communications
Management and Manage Communications
processes; C) focuses on acquiring, training, and
releasing staff, along with recognition programs,
compliance, and safety; D) resolved issues are
more likely to be outputs of the Control
Communications process.

15. C Section 10.3 – Monitoring and Controlling
A) is part of the Control Stakeholder Engagement
process; B) is part of the Plan Communications
Management process; D) is part of the Plan
Stakeholder Management process.

10

CASE STUDY SUGGESTED SOLUTIONS

Exercise 10-1

$[50 \, (50–1)] \div 2 = 2450 \div 2 = 1225$

Exercise 10-2

Communications Matrix for the Lawrence Garage Project

The project manager will produce all required communications as stated below.

Communications Matrix								
	Report Progress Weekly	Weekly Status Meeting	Weekly Earned Value Reports	Phase Review Meeting	Phase-end Earned Value Reports	Final Project EV Reports	Final Project Acceptance	Lessons Learned Report
Owner		A	RR	A	RR	RR	A	RR
Architect		A	RR	A	RR	RR	A	RR
General contractor	R	A	P	A	P	P	A	P
Site excavation crew	R	A*						
Concrete crew	R	A*						
Framing crew	R	A*						
Drywall crew	R	A*						
Painting crew	R	A*						
Finish carpentry crew	R	A*						
Electric contractor	R	A*	RR	A*	RR	RR		RR
Plumbing contractor	R	A*	RR	A*	RR	RR		RR
Roofing contractor	R	A*	RR	A*	RR	RR		RR
Insulation contractor	R	A*	RR	A*	RR	RR		RR
Heating and AC contractor	R	A*	RR	A*	RR	RR		RR
Lath & plaster contractor	R	A*	RR	A*	RR	RR		RR
Inspector							A*	
Electric company							A*	

R = Report progress
P = Prepare report
A = Attend meeting (* = when appropriate)
RR = Review report

RISK

CHAPTER 11 | **RISK**

11

RISK MANAGEMENT

Project risk management is considered by some to be the most difficult section of the PMP exam. Exam takers consider it demanding because it addresses many concepts that project managers have not been exposed to in their work or education. Questions, however, do correspond closely to *PMBOK® Guide* material, so you should not have much difficulty if you study the terminology found in this guide.

The mathematical questions are not very difficult; however, they do require you to know certain theories, such as **expected monetary value** and **decision tree analysis**. You should also expect questions related to levels of risk faced by both buyer and seller based on various types of contracts.

Project managers have an ethical responsibility to communicate potential risks and their impact on a project's success.

> **EXAM TIP**
> Project risk can be decreased significantly by actively planning, identifying, assessing, and developing responses to potential risk events and managing their impact on a project.

Things to Know

1. The six processes of risk management:
    *   **Plan Risk Management**
    *   **Identify Risks**
    *   **Perform Qualitative Risk Analysis**
    *   **Perform Quantitative Risk Analysis**
    *   **Plan Risk Responses**
    *   **Control Risks**
2. The **risk management plan**
3. The differences between **risk appetite**, **risk tolerance**, and **risk threshold**
4. The **utility theory**
5. **Risk categories** and their uses, as well as the **risk breakdown structure**
6. The tools and techniques for identifying risks:
    *   **Brainstorming**
    *   **Delphi technique**
    *   **Interviewing**
    *   **Root cause analysis**
    *   **SWOT analysis**
    *   **The risk register**

11

7. Risk **probability and impact**
8. Quantitative analysis tools of:
   - **Interviewing**
   - **Probability distributions**
   - **Decision tree**
   - **Expected monetary value analysis**
   - **Sensitivity analysis**
   - **Monte Carlo simulation**
9. **Risk response strategies**
10. Key **interpersonal skills** for success

Key Definitions

Contingency plan: a planned response to a risk event that will be implemented only if the risk event occurs.

Contingency reserve: a dollar or time value that is added to the project schedule or budget and that reflects and accounts for risk that is anticipated for a project.

Decision theory: a technique for assisting in reaching decisions under uncertainty and risk. It points to the best possible course, whether or not the forecasts are accurate.

Fallback plan: a response plan that will be implemented if the primary response plan is ineffective.

Heuristics: rules of thumb for accomplishing tasks. Heuristics are easy and intuitive ways to deal with uncertain situations; however, they tend to result in probability assessments that are biased.

Issue: a risk event that has occurred.

Opportunities: risk events or conditions that are favorable to the project.

Residual risk: when implementing a risk response plan, the risk that cannot be eliminated.

Risk: an uncertain event or condition that could have a positive or negative impact on a project's objectives. Therefore, the primary elements of risk that must be determined are:
- Probability of the risk event or condition occurring
- Impact of the occurrence, if it does occur
- Expected time the risk event may occur
- Anticipated frequency of the risk event occurring

Risk appetite: the degree of uncertainty an entity is willing to take on in anticipation of a reward.

Risk threshold: the measures, along the level of uncertainty or the level of impact, at which a stakeholder may have a specific interest. Risk will be tolerated under the threshold and not tolerated over the threshold.

Risk tolerance: the degree, amount, or volume of risk that an organization or individual will withstand.

Secondary risk: when implementing a risk response, a new risk that is introduced as a result of the response.

Threat: risk events or conditions that are unfavorable to a project.

Workarounds: unplanned responses to risks that were previously unidentified or accepted.

PLAN RISK MANAGEMENT PROCESS

The Plan Risk Management process plans for risks that may occur during a project. It is the process of deciding how to approach and plan for activities to handle project risks and documenting these decisions in a **risk management plan**, which is the primary output of the Risk Management Planning process. This important document is a subsidiary component of the project management plan.

EXAM TIP

The risk management plan contains the:
1. Risk approach and methodology
2. Roles and responsibilities of the risk management team
3. Risk management budget
4. Timing of the risk management process
5. Risk categories
6. Definitions of probability and impact
7. Probability and impact matrix
8. Revised stakeholder risk tolerances
9. Risk planning report formats
10. Methods of tracking risks

Key inputs to this process are the **stakeholder register**, the **project scope statement**, the schedule, cost and communications management plans, as well as the **enterprise environmental factors** and **organizational process assets**.

The organizational **culture** and stakeholders' attitudes towards risk are factors that must be considered as part of the project environment. Organizational and individual tolerances for risk are not often considered an aspect of project management, but they will make a difference in the approach taken toward project risks. Since different organizations and individuals have varying levels of tolerance for risk, decisions should be made based on the tolerances of the project team.

And a single individual may have different risk tolerance levels when approaching different projects. An entrepreneur may have a high **appetite** for risk, but depending upon the venture, the **threshold** and **tolerance** can change. For example, an entrepreneur manufacturing a new medical device will have a very different risk profile than an entrepreneur manufacturing a new golf club. To assess an organization's willingness to take on risk, you have to consider the appetite, tolerance, and threshold for risk for that organization.

EXAM TIP

When presented with a probability theory problem, come up with all possibilities and verify that they all sum up to "1" to avoid simple errors in math. For example, if the probability of rain is 10%, then the probability of no rain is 90%.

11

Utility Theory

An appropriate method for describing **risk tolerance** is the **utility theory**.

Figure 11-1 below depicts the three structures of the utility theory. The x-axis denotes the money at stake and the y-axis denotes utility, or the amount of satisfaction a person obtains from the payoff.

There are three types of preferences to consider when considering stakeholder risk. They are:
- **Risk averse**: when there is more money at stake, the risk averter's satisfaction diminishes; he or she prefers a more certain outcome and demands a premium to accept projects of high risk
- **Risk neutral**: tolerance for risk remains the same as the money at stake increases
- **Risk seeker**: for the risk seeker, the higher the stakes, the better; as risk increases, the risk seeker's satisfaction increases; he or she is even willing to pay a penalty to take on projects of high risk

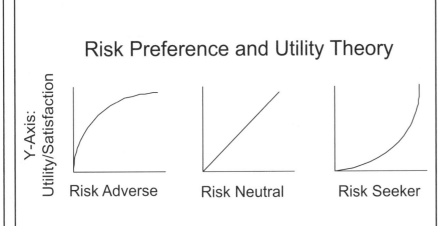

Risk Preference and Utility Theory

Y-Axis: Utility/Satisfaction

Risk Adverse   Risk Neutral   Risk Seeker

Figure 11-1
Risk Preference and
the Utility Theory
(from Kerzner's
*Project Management:
A Systems Approach to
Planning, Scheduling,
and Controlling*, page
655)

11

Risk Categories

Risk categories are commonly included in the **risk management plan**. An organization may have a standard set of risk sources that should be considered in advance of the Identify Risks process. These sources of risk are defined in the **risk breakdown structure** (RBS). There may be many categories of risk, including:

- **External risks**, such as from vendors, regulations, customers, and market conditions; for example, you know that snow storms will occur in the northwest, but you do not know when and how many snow storms will occur (note that *force majeure* **risks**, such as earthquakes, floods, acts of terrorism, etc., should be covered under disaster recovery procedures instead of risk management)
- **Project management risks**, such as poor estimates of time and resources or the lack of skills and knowledge in project management concepts and discipline by the project manager and/or team members
- **Organizational risks**, such as resource conflicts, a delay in the availability of resources, or a lack of funds
- **Technical risks**, such as complex or new software technology, quality or performance issues, or unrealistic project goals

The **risk probability and impact matrix** is another component of the risk management plan. The risk probability and impact matrix will be **tailored** to the project based on the project's needs.

IDENTIFY RISKS PROCESS

The Identify Risks process involves identifying and documenting the types of risks that may occur during a project. It is important to think of each risk event as having a three-part anatomy:

- The uncertain event or condition that poses the risk
- The impact of that event or situation
- The source of the risk

The identification of risks is an iterative process that occurs with the help of the project team, stakeholders, and even people outside the **organization**. **Information gathering** techniques are used in this process. Information can be gathered by many means, including **brainstorming**, the **Delphi technique**, **interviewing**, and a **strengths**, **weaknesses**, **opportunities**, and **threats** (SWOT) analysis.

Brainstorming

Information gathered through brainstorming is:

- Used extensively in project planning
- Possibly used to postulate risk scenarios for a particular project
- Improved by including participants with a variety of backgrounds
- Helpful in project team building
- Effective in finding solutions to potential problems

The Delphi Technique

The **Delphi technique** is a form of expert judgment in which opinions are obtained from a panel of experts who work independently and anonymously. It is often used in risk management but can also be used to gain consensus on project selection, scope of work, estimates, and technical issues. The Delphi technique:

- Derives a consensus using a panel of experts to arrive at a convergent solution to a specific problem

11

- Is useful in arriving at probability assessments relating to risk events in which the risk impacts are large and critical

Interviewing

Experienced project managers or subject matter experts are interviewed to identify project risks based on their knowledge of technical or external issues.

Root Cause Analysis

Problems are often not easily identifiable. Using **root cause analysis** techniques, the project manager can discover the underlying causes that lead to a problem and therefore develop preventive actions.

SWOT Analysis

SWOT analysis is a technique that:
- Examines potential risks from the perspectives of strengths, weaknesses, opportunities and threats, i.e., SWOT, to look for internally generated risk events

Risk Register

The only output of the Identify Risks process is the **risk register** that contains a list of the identified risks and potential response strategies. The risk register is used to document risk planning and assessment activities and to track the status, triggers, and responses to project risks. The identification, tracking, and reviewing of project risks is ongoing throughout the **project life cycle**. The risk register is a component of the risk management plan and the project management plan. It is updated in each of the risk management processes. An example of a risk register is shown in Figure 11-2 on the next page.

EXAM TIP

The risk register is a critical deliverable in any project. Every process within the risk management knowledge area includes updates to the risk register as a process output, reinforcing the concept of an iterative approach to risk planning.

11

Risk Events	Category	Probability	Impact	Score/ Priority	Risk Response Strategy	Risk Trigger
Poor Estimate of Testing	Project Management	2	4	8	Mitigate	Missed Milestone
Part Failure	Technical	2	2	4	Transfer	Test Failure
PM Promoted	Organization	3	4	12	Accept	Reorgani- zation
Shipping Delay	External	4	5	20	Mitigate	Labor Dispute

Figure 11-2
Sample Risk Register

PERFORM QUALITATIVE RISK ANALYSIS PROCESS

Once risks have been identified, they must be analyzed to determine the likelihood of the risk occurring (**risk probability**) and the consequences it could have on the project (**risk impact**) if and when it occurs. This process has an advantage of being a rapid and cost effective way to assess risks.

Qualitative risk analysis involves the activities of:
- Assessing the probability and impact of identified risks
- Determining timing and urgency
- Prioritizing risks
- Ranking risk events in order of importance
- Evaluating the quality of data related to each risk
- Determining which risks require additional analysis

Risk Probability and Impact

The **risk management plan** should define the **probability and impact scale** to be used in assessing and determining the relative importance of each risk for the project. When calculating the **risk score**, which is the product of the probability times the impact, the project team will determine if the risk is a low, moderate, or high priority.

> **EXAM TIP**
> The probably and impact matrix includes the scale that was defined by the organization. The probability and impact matrix is an easy and effective way for the project manager to access the priority of issues on a project.

11

Probability & Impact Score for a Risk					
Probability	Risk Score = P X I				
5	5	10	15	20	25
4	4	8	12	16	20
3	3	6	9	12	15
2	2	4	6	8	10
1	1	2	3	4	5
	1	2	3	4	5
	Impact (Ratio Scale)				

HIGH

MED

LOW

Figure 11-3
Risk Probability
and Impact Matrix

For example, you have a year-long project in northern Maine and Figure 11-3 was provided as part of project planning. You now have determined that snow could be a risk that could impact the schedule. The likelihood that a snowstorm will occur in Maine this winter is high. But since our workforce works virtually, the impact of a large storm on workforce productivity is low. This risk event would have a score of 5 (high probability) X 1 (low impact) for a score of 5. It would be reflected in the upper left hand corner of the grid in Figure 11-3 and considered a medium risk.

PERFORM QUANTITATIVE RISK ANALYSIS PROCESS

Quantitative risk analysis involves further analyzing identified risk events for their effect on **project objectives**. This analysis may include:
- Numerically analyzing the effects or probability of achieving objectives, often stating the effects in monetary terms

- Evaluating the range of impacts of risks on the project objectives of cost, schedule, scope, and quality
- Determining the extent of overall project risk
- Using probabilistic models to determine trends, ranges of acceptable risk, and probability of success for a given value

Interviewing

Just as interviews with subject matter experts and other stakeholders are useful in identifying risks, they are also important in quantifying risks and defining ranges of values. These values typically take the form of three-point estimates for cost or duration, which are used in revising budget or time estimates.

Probability Distributions

Various types of probability distributions are used to represent the uncertainty that risks represent. Discrete distributions may be used in developing decision tree values. Continuous distributions, such as the beta or triangular distributions shown in Chapter 1, are used in modeling and simulations.

Decision Tree

A decision tree is used when a choice needs to be made among several options, and those options are affected by several key variables. The decision tree diagram depicts key interactions among risk events, and it shows:

- The outcomes, typically described in monetary terms
- The risk events, related to time, cost, scope, or quality considerations
- Probability and impact in each branch of the decision tree
- Expected monetary value for each individual path

11

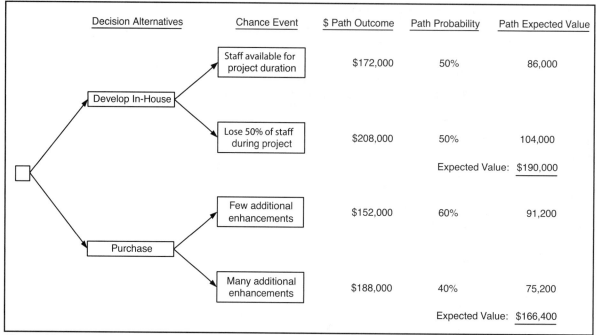

Figure 11-4
A Decision Tree

Figure 11-4 above shows a decision tree analysis for the choice of developing software in-house versus purchasing software. The risk if the company develops software in-house is that it could lose 50% of its staff during the project. This loss would cause the cost of the project to increase from $172,000 to $208,000 because contract labor will be required.

The risk for purchased software is the uncertain number of enhancements that will be necessary for it to be implemented. If there are few enhancements, the cost will be $152,000. If there are many enhancements, the cost of the project will be $188,000.

Based on the identified risks, the expected cost for developing in-house is $190,000 while for purchasing it is $166,400. Purchasing the software is the better choice to minimize the overall project cost.

Expected Monetary Value Analysis

A risk event may be quantified by combining probability and impact, and this quantification is the expected monetary value (EMV).
- For opportunities, the EMV will be a positive value
- For threats, the EMV will be a negative value
- EMV is the basis for most simulation programs
- EMV provides a predictor of the outcome that has the least bias
- EMV is used with the discrete risk events depicted in a decision tree

Sensitivity Analysis

Sensitivity analysis is a modeling technique used to compare the relative importance of variables. It may be applied to different risk events, with the most sensitive variable being the one that has the greatest influence on a project. The implication is that the project team should focus on the risks with the greatest sensitivity.

Monte Carlo Simulation

With the advent of specialized software add-ins to most project management software packages, a Monte Carlo simulation has become an accessible and powerful technique to model a project and its associated risks. The model includes the following:
- A range of outcomes for each variable
- A specified probability distribution
- A random number generation of the results of the project activities based on the specified range and probability distribution
- A large number of iterations of the project activities

EXAM TIP

After risk analysis, the scope, schedule, and/or cost baselines may need to be updated to reflect new or omitted work. This is NOT considered scope creep.

11

The results of a Monte Carlo simulation provide a cumulative probability distribution that shows the likelihood of achieving a particular value of cost or time. It helps the project team determine the amount of contingency that would be appropriate for the project.

PLAN RISK RESPONSES PROCESS

During the Plan Risk Responses process, the action plans for how risks should be handled are determined. This process includes the activities of:

- Developing options
- Determining actions
- Enhancing opportunities
- Reducing threats
- Assigning responsibility for risks
- Adding resources or activities to the project management plan
- Determining appropriate responses based on the priority of risks

The primary outputs of this process are the updates to the **risk register**, updates to the project management plan, and any risk-related contractual **agreements**.

Risk Response Strategies

The tools and techniques for risk response planning have been divided into strategies for threats, opportunities, and contingent response strategies.

The strategy that can be used for both **threats** and **opportunities** is:

- **Accept**: acceptance means accepting the consequences of the risk; acceptance can be active, such as developing a contingency plan should the risk occur, or it can be passive, such as accepting a lower profit if some activities overrun

The strategies that can be used for **negative risks** or **threats** are:

- **Avoid**: avoidance eliminates a specific threat, usually by eliminating the cause; examples of avoidance include not doing a project or doing the project in a different way so that the risk no longer exists

---

EXAM TIP

An "accept" strategy is typically used for risks with a low probability and impact score.

11

- **Transfer**: transference reduces the direct risk by shifting it to a third party, such as insurance companies, clients, or vendors; transferring does not eliminate the overall risk
- **Mitigate**: mitigation reduces the EMV of a risk event by reducing the probability of its occurrence or reducing the impact of the risk; an example of mitigation would be using proven technology to lessen the probability that a product will not work

The strategies that can be used for **positive risks** or **opportunities** are:

- **Exploit**: exploitation increases the chances of making an opportunity happen, and it helps eliminate the uncertainty
- **Share**: sharing is done with a third party, and leverages that party's ability to capitalize on the opportunity
- **Enhance**: enhancing increases the probability of an opportunity's occurrence or its positive impact

> **EXAM TIP**
> You can plan contingencies for any project constraint, such as time or resources.

Contingent response strategies are designed to address specific situations if a certain event occurs. These differ from known risks or opportunities for your project described above. For example, every project could theoretically be impacted due to a fire in the office. You wouldn't necessarily put that in your risk register. Your response would be part of a contingent plan or fall back plan that would be triggered in the unlikely event that a fire occurs. Most likely your organization already has a plan for the handling of a fire situation.

Case Study Exercise

Exercise 11-1: Create a risk management plan for the Lawrence Garage Project. Identify at least three potential risk events for the Lawrence Garage Project. Perform qualitative analysis on those risk events and develop a response plan for each of the events. Create a risk register summarizing your risk response planning.

CONTROL RISKS PROCESS

The Control Risks process involves the tracking of identified risks that have or have not yet occurred, identifying new risks, and recalculating risk scores for increased or decreased priority. Thus, the iterative steps of this process are to determine if:

- Risk responses have been implemented as planned
- Risk responses are effective
- Project assumptions are still valid
- Risk exposure has changed
- Risk triggers have occurred
- Policies and procedures have been followed
- New risks have been identified
- Risk attitudes, tolerances, or thresholds have not changed
- Risk events have occurred that were previously not defined

Variance and trend analysis are techniques of the Control Risks process. **Earned value management** and other techniques may be used to monitor project performance. If these calculations show a significant deviation from cost and schedule baselines. Earned value management is discussed in depth in Chapters 6 (Time) and 7 (Cost). Some tips for controlling risks are:

- Work the top priority risk events
- Appoint a risk response owner for top risks
- Review performance weekly, as earned value analysis may indicate deviations to investigate
- Work all top risks concurrently
- Reprioritize the top risks as time passes and as new risks are identified and added
- Take **corrective actions** or use **contingency plans** as necessary
- Continue tracking and adjusting at regular intervals throughout a project's life cycle
- Use **workarounds** for unplanned events to minimize loss, repair damage, and prevent the recurrence of a risk event

11

Risk audits can also be an invaluable tool to assess the effectiveness of risk response strategies and the overall risk management processes' effectiveness.

11

KEY INTERPERSONAL SKILLS FOR SUCCESS

The interpersonal skill highlighted in this chapter is:

Influencing

Project managers need to build strong influencing skills since they frequently lack legitimate authority to manage team members. Influencing means the project manager gains the cooperation of stakeholders without traditional reward and penalty motivations. Project managers often accomplish this by practicing fairness, honesty, and respect, and by taking responsibility.

When evaluating risks, people will have differing views of the probability and impact of each risk. Keen influencing skills can help others gain consensus to move the project forward and not get stuck in "analysis paralysis."

SAMPLE PMP EXAM QUESTIONS ON RISK MANAGEMENT

1. You are the project manager for a business process improvement project for a strategic business process, and the department responsible for a deliverable on the critical path is running behind plan. The department manager has not been responding to your phone calls and is unavailable for meetings. A member of his team has told you there is a plan to bring the deliverable back on schedule, but you haven't seen it. You should:

   a) Identify the risk in the risk register and monitor progress more closely over the upcoming week
   b) Create an activity in the WBS
   c) Add two weeks to the activity duration as a contingency plan
   d) Storm into the department manager's office and demand the plan from him

2. Risk audits can identify:

   a) The effectiveness of the risk management process
   b) Estimated project durations for completed activities
   c) The basis for planning when associated work packages have not yet been planned
   d) The amount of contingency reserves needed

3. _____ are unplanned responses to threats that were previously accepted or not identified.

   a) Workarounds
   b) Expenses
   c) Contract changes
   d) Contingency plans

4. Emma, a project manager at a large utility company, developed a contingency plan for a risk event on her project. She was unable to implement the contingency plan and had to rely on a:

   a) Residual plan
   b) Fallback plan
   c) Management plan
   d) Assessment plan

5. A charity event party is to be held that will cost $10 per person. Here is the information you have:

   If the party is held inside, 150 people are likely to attend
   If the party is held outside, 200 people are likely to attend
   There is a 20% chance of rain on the date of the party. If it rains and the party is held inside, only 100 people will attend.
   If it rains and the party is held outside, only 80 people will attend.

   In this example, if the goal of the project is to maximize revenue, should the party be held inside or outside, and why?

   a) Held inside because expected monetary value is $1,200
   b) Held inside because expected monetary value is $1,400
   c) Held outside because the expected monetary value is $1,400
   d) Held outside because the expected monetary value is $1,760

6. Which of the following are tools and techniques of the Plan Risk Management process?

   a) Checklist analysis and assumptions analysis
   b) Monte Carlo simulation and decision tree
   c) Meetings and expert judgment
   d) Probability/impact matrix and risk categories

Notes:

7. Which of the following statements is true regarding project risk?

   a) Risks are primarily related to negative impacts on at least one project objective
   b) Project risk is something that has already happened
   c) Risks may have positive or negative outcomes
   d) Project risk should include the chance of a unique event such as a tsunami

8. Isabella is the project manager on a software development project that is using a state-of-the-art programming language. She has identified human resources risks to have the highest probability of occurrence and highest impact if they occur. Which of the following would be the least appropriate response?

   a) Change the scope of the project to use a programming language you are more accustomed to, which is COBOL
   b) Provide additional training to team members in the expertise that is most in need
   c) Bring a few contractors onto the team who have proven experience in the desired knowledge
   d) Review the scope to verify the use of the state-of-the-art programming language and focus your limited resources there

9. Known unknown risks above your priority threshold that are not managed proactively should have a:

   a) Management reserve
   b) Reserve analysis
   c) Mitigation strategy
   d) Contingency reserve

Notes:

11

10. The overall perspective on the level of risk that an individual or organization is willing to take is called the:

   a) Risk attitude
   b) Risk appetite
   c) Risk tolerance
   d) Risk threshold

11. A technique for risk monitoring and control is:

   a) Expert judgment
   b) Risk categorization
   c) Risk urgency assessment
   d) Risk reassessment

12. When managing a project which has multiple phases, it is likely that risks will:

   a) Be different in each phase
   b) Stay consistent from phase to phase
   c) Be resolved at the end of a phase
   d) Have residual risks

13. A sensitivity analysis will help you determine which risks have:

   a) The highest probability
   b) The most potential impact on the project
   c) A specific root cause
   d) The lowest impact on the project

Notes:

11

14. Ava identified a significant number of risks and used a probability and impact matrix to focus on the most important risk events. She wanted to conduct a quantitative analysis but was unable to do so. What is the most likely reason?

    a) There were too many identified risk events to be included in a model
    b) There was no provision for quantitative analysis in the risk management plan
    c) There was insufficient data that was appropriate for a model
    d) The budget for contingencies did not include a line item for a model

15. As the project manager on a software development project, you find that your team has identified that the most important risk for your project is the ability to retain technical expertise on the project. Without any risk strategy, if the risk event occurs:

    a) There can only be an impact to the schedule
    b) There will be an impact to both the schedule and the quality of the project
    c) There will only be an impact to the cost of the project
    d) There will not be any impact to the project

Notes:

11

ANSWERS AND REFERENCES FOR SAMPLE PMP EXAM QUESTIONS ON RISK MANAGEMENT
Section numbers refer to the *PMBOK® Guide*.

1. **A** **Section 11.6 – Monitoring and Controlling**
    B) the activity should already exist in the WBS;
    C) adding two weeks to the activity may be
    premature and/or inadequate.

2. **A** **Section 11.6 – Monitoring and Controlling**
    B) is a combination of two competing activities:
    estimating is completed during the Estimate
    Activity Durations process and documenting
    completed durations takes place during the
    Control Schedule process; C) is usually part of
    rolling wave planning, and a risk audit will have
    minimal or no input to this work; D) the risk audit
    is better used to determine the amount and
    effectiveness of contingency reserves used, not
    needed.

3. **A** **Section 11.6 – Monitoring and Controlling**
    Workaround plans must be properly documented
    and incorporated into the project plan and the risk
    response plan.

4. **B** **Section 11.5 – Planning**
    A) a residual plan may be developed for the risk
    remaining after the initial risk response plan is
    developed; C) and D) are too general and not
    specific to project risk

5. **D** **Section 11.4 – Planning**
    Held outside because the expected monetary value
    for outside is $1,760 which is greater than $1,400
    if it is held inside. There are only two choices:
    hold inside or hold outside. You cannot choose
    whether it rains or not.

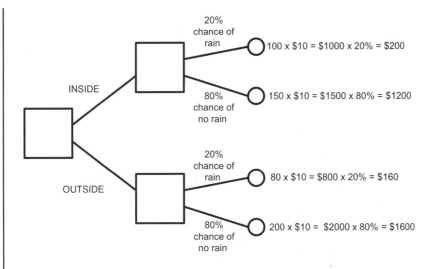

6. **C Section 11.0 – Planning**
   A) belong to the Identify Risks process; B) belong
   to the Perform Quantitative Risk Analysis process;
   D) belong to the Perform Qualitative Risk Analysis
   process.

7. **C Section 11.5 – Planning**
   A) risk management should address both positive
   and negative risks; B) issues are risks that have
   happened; D) catastrophic events are handled at
   the organizational level, not the project level.

8. **A Section 11.0 – Planning**
   Resorting back to an old architecture tool is not
   necessarily the best thing to do.

9. **D Section 11.0 – Planning**
   Management reserves are used for unknown
   unknown risks.

10. **A Section 11.6 – Monitoring and Controlling**
    B) risk appetite is the degree of uncertainty offset
    by the anticipation of a reward; C) risk tolerance is
    the amount of risk someone is willing to accept; D)
    risk threshold is the dividing line between accept-
    ing or not accepting risk.

11

11. D **Section 11.2 – Planning**
A) is a technique in all the risk planning processes; B) risk appetite is a measure of risk that an organization is willing to take on; C) risk tolerance is the level of risk that an organization will stand.

12. A **Section 11.2 – Planning**
Risk continually changes throughout a project due to changes in situations. Risks must be regularly reviewed and discussed to ensure proactive attention.

13. B **Section 11.4 – Planning**
A) and D) probability and impact must be assessed before performing a sensitivity analysis; C) an Ishikawa or fishbone diagram is used to determine causes, but doesn't determine which cause has the greatest effect on a project.

14. C **Section 11.4 – Planning**
A) models can generally handle a large number of items; B) may be true, but C) is a better answer; D) the contingency budget should have no effect on the costs for a model.

15. B **Section 11.3 – Planning**
Technical expertise can have an impact on quality, and any time you have resource issues there will most likely be an impact to the schedule.

CASE STUDY SUGGESTED SOLUTION

Exercise 11-1
Risk Management Plan for the Lawrence Garage Project

Qualitative estimates will vary greatly depending on personal experience. The key is to think about risks BEFORE they happen and to have a plan for dealing with risks. Although we have indicated a possible response plan and risk register for only one of the many possible risks, a project manager would have to address all valid risks to a project.

Some possible risk events:
- Permit delays
- Permit(s) not granted
- Subcontractor bids over (or under) expectations
- Schedule overruns
- Weather delays
- Materials delays
- Labor delays
- Schedule delays other than weather, materials, or labor
- Budget overruns due to materials, labor, or subcontractor invoices
- Owner or architect change requests
- Inspection denials
- Quality problems

Response plan for permit delays:
As a mitigator, review the permit applications to determine that they have been completed properly with all the necessary attachments. As a contingency plan, if a permit is delayed, meet with the permitting agency to determine the cause of the delay and the steps to extradite the permit.

Risk register for permit delays:

Risk Event	Category	Probability	Impact	Score/ Priority	Risk Response Strategy	Risk Trigger
Permit Delayed	External	3	2	6	Accept	Permit not Received

PROCUREMENT

CHAPTER 12 | **PROCUREMENT**

PROCUREMENT MANAGEMENT

Procurement management questions on the PMP exam tend to be more process oriented than legally oriented. You do not need to know any country's specific legal code; however, the nature of many of the questions requires an understanding of United States contract law. In the United States, a contract is a formal agreement, all changes must be in writing and formally controlled, and a court system is used for handling disputes. You must have a firm understanding of the procurement process in order to answer these questions accurately.

You must know the basic differences between the primary categories of contracts (cost reimbursable, fixed price and time and material) and the risks inherent in each category for both the buyer and the seller.

Several questions will also test your knowledge of various contract types within each category. **International contracting** is within the limits of exam questions; the timing of foreign currency exchange and duty on goods delivered to foreign countries may also come up.

> **EXAM TIP**
> The *PMBOK® Guide* discusses procurement management from the perspective of the buyer in the buyer-seller relationship, and from the perspective that the seller is external to the buyer's organization. The type, terms and conditions of the contract become key inputs in the contract administration process.

Things to Know

1. The four processes of procurement management:
   - **Plan Procurement**
   - **Conduct Procurements**
   - **Control Procurements**
   - **Close Procurements**
2. **Teaming agreements**
3. The three contract types:
   - **Fixed price**
   - **Cost reimbursable**
   - **Time and material**
4. The purpose of the **make-or-buy analysis**
5. The benefits of **market research**
6. The **procurement management plan**
7. The procurement **statement of work (SOW)**
8. Types of **specifications**

9. Types of procurement documents:
    *   **Invitation for bid**
    *   **Request for bid**
    *   **Request for quotation**
10. Evaluating sellers with **source selection criteria**
11. Procurement **negotiation**
12. **Negotiation tactics**
13. How to define **agreements**
14. What the **contract change control system** is
15. Challenges in **closing procurements**
16. **Closure outputs**
17. Key **interpersonal skills** for success

Key Definitions

Bidder conference: the buyer and potential sellers meet prior to the contract award to answer questions and clarify requirements; the intent is for all sellers to have equal access to the same information.

Buyer: the performing organization, client, customer, contractor, purchaser, or requester seeking to acquire goods and services from an external entity (the seller). The buyer becomes the customer and key stakeholder.

Commercial-off-the-shelf (COTS): a product or service that is readily available from many sources; selection of a seller is primarily driven by price.

Contract: the binding agreement between the buyer and the seller.

Letter contract: a written preliminary contract authorizing the seller to begin work immediately; it is often used for small value contracts.

Letter of intent: this is NOT a contract but simply a letter, without legal binding, that says the buyer intends to hire the seller.

Point of total assumption: in a fixed price contract, the point above which the seller will assume responsibility for all costs; it generally occurs when the contract ceiling price has been exceeded.

Privity: the contractual relationship between the two parties of a contract. If party A contracts with party B, and party B subcontracts to party C, there is no privity between party A and party C.

Seller: the bidder, contractor, source, subcontractor, vendor, or supplier who will provide the goods and services to the buyer. The seller generally manages the work as a project, utilizing all processes and knowledge areas of project management.

Single source: selecting a seller without competition. This may be appropriate if there is an emergency or prior business relationship.

Sole source: selecting a seller because it is the only provider of a needed product or service.

PLAN PROCUREMENT MANAGEMENT PROCESS

In the Plan Procurement Management process, the project team identifies which **project requirements** can be met by purchasing products or services from sellers outside the organization. It will include documenting decisions to buy and the process to procure products or services. The team, along with the procurement department, will also identify **sellers** and special circumstances that may influence the sellers who may bid. The Plan Procurement Management process should take place during the Define Scope process and will rely on the **scope statement**, **WBS**, **documentation of requirements**, **risk register**, and other planning outputs related to the schedule, resource requirements, quality, and cost. Most organizations have some level of formal purchasing procedures that must be followed.

EXAM TIP

Apart from the Plan Procurement Management process, the procurement processes are the only *PMBOK® Guide* processes that are truly optional. These processes are only used if procurement is determined to be necessary to the project.

12

Contract Types

The primary objective in selecting a type of contract is to have risk distributed between the buyer and seller so that both parties have **motivation** and **incentive** to meet the contract goal. If a product or service is not well defined, both the buyer and seller are at risk.

The following factors may influence the type of contract selected:
- Overall degree of cost and schedule risk
- Type/complexity of requirement
- Extent of price competition
- Cost/price analysis
- Urgency of requirement/performance period
- The level of detail in the statement of work
- Frequency of expected changes
- Industry standards for types of contracts used

There are three general types of contracts:

- **Fixed price** (FP) or lump sum contracts involve a predetermined fixed price for a product; it is the preferred type of contract in many situations and is used when the product is well defined
- **Cost reimbursable** (CR) or **cost plus** contracts involve payment based on the seller's actual costs, a fee, and a potential incentive for meeting or exceeding project objectives; it may be easier to initiate but requires more oversight by the buyer
- **Time and material** (T&M) contracts (sometimes called **unit price** contracts) contain characteristics of both FP and CR contracts and are generally used for small dollar amounts; these contracts may be priced on a per-hour or per-item basis (like fixed price contracts) but the total number of hours or items is not determined (open-ended cost-type arrangements like CR contracts)

Fixed Price Contracts

In fixed price contracts, the seller is paid an agreed-upon price for the work to be performed which must include the seller's profit and any necessary contingencies. Therefore, the seller bears a higher burden of the cost risk than the buyer. Typical types of fixed price contracts are:

- **Firm-fixed price** (FFP): also called a lump sum contract, in which the seller bears the greatest degree of risk; this is a common type of contract when the seller agrees to perform a service or furnish supplies at an established contract price

  One example of an FFP contract:

Contract price	$100,000
Seller's actual cost	$70,000
Buyer's total payments	$100,000

  The seller's actual cost is usually not disclosed in this type of contract; the seller's profit is $30,000 in this example

  Another example of an FFP contract:

Contract price	$100,000
Seller's actual cost	$130,000
Buyer's total payment	$100,000

  The seller loses $30,000 on this contract

- **Fixed price incentive fee** (FPIF): a complex type of contract in which the seller bears a higher burden of risk, but the purpose of the incentive is to shift some of the risk back to the buyer; typically, for every dollar the seller reduces cost below the target, the cost savings are split between the buyer and seller based on the share ratio (a ceiling price is established so the buyer does not pay more than the ceiling price; therefore, if costs exceed the ceiling, the seller receives no additional payments)

12

One example of an FPIF contract:

Target cost	$100,000
Target profit	$10,000
Contract price value	$110,000
Ceiling price	$120,000
Share ratio (buyer/seller)	80-20
Seller's actual cost	$80,000
Difference	$20,000

Profit calculation:

Seller's share of difference (20%)	$4,000
Target profit	$10,000
Total profit	$14,000

Total calculation:

Actual cost	$80,000
Total profit	$14,000
Buyer's total payment	$94,000

Another example of an FPIF contract: same targets and ceiling as above; if the project's actual cost increases to $130,000, the buyer pays the seller $120,000 (the ceiling amount) and the seller receives no additional payments, which means the seller may have no profit on the product or service sold

- **Fixed price with economic price adjustment (FP-EPA):** this contract may be used when the term of the contract spans multiple years; it provides for adjustment to the contract prices when specified economic events occur and is intended to reduce the seller's risk of price changes of purchased items

  An example of an FP-EPA contract: contract price is $100,000 for each of three years; if a commodity used in the production of the product or service increases due to inflation, the contract price will be increased

Cost Reimbursable Contracts

In this type of contract, the buyer and seller agree to the costs to be reimbursed and to the amount of profit. Therefore, the buyer bears the highest cost risk. Common forms of cost reimbursable contracts include:

- **Cost plus fixed fee** (CPFF): this is a common form of cost reimbursable contract in which the buyer bears the burden of the cost risk because the buyer pays all costs; the seller's fee, or profit, is fixed at a specific dollar amount and may be a percent of the original estimated cost; this type of contract is often used in research projects and projects in which the scope of work lacks clear definition

  One example of an CPFF contract:

Estimated cost	$100,000
Fixed fee	$10,000
Contract value	$110,000
Seller's actual cost	$150,000
Fixed fee	$10,000
Buyer's total payment	$160,000

  Another example of a CPFF contract may be when the fee is renegotiated at the time that a buyer approves changes that increase or decrease the scope of the project.

Estimated cost:	$100,000
Fixed Fee:	$10,000
Contract value:	$110,000
Approved scope changes of	$20,000
New Estiamted cost:	$120,000
Renegotiated Fee:	$12,000
New Contract value:	$132,000

12

EXAM TIP

Buyers' risk from the various contract types (from highest to lowest):

CPFF > CPIF > FPIF > FFP

- **Cost plus incentive fee** (CPIF): in this type of contract, risk is shared by the both buyer and seller, and the seller is paid for agreed to, allowable costs plus an agreed upon fee, and an incentive that is both specific and measurable; if the final costs are less than the expected costs, both the buyer and seller benefit by splitting the cost savings on a prenegotiated sharing formula; this type of contract is used for long performance periods.

  The difference between FPIF and CPIF contracts is that FPIF contracts include a ceiling, above which the seller will not recover any cost; CPIF contracts do not have a ceiling on cost, but they often have maximum and minimum fees

  An example of a CPIF contract:

Target cost	$100,000
Target fee	$10,000
Maximum fee	$15,000
Minimum fee	$7,000
Share ratio (buyer/seller)	80-20
Seller's actual cost	$120,000
Difference (over target cost)	($20,000)

  Profit calculation:

Target profit	$10,000
Seller's share of overage (20%)	$4,000
Calculated profit	$6,000
Minimum profit	$7,000

  Total calculation:

Actual cost	$120,000
Total profit	$7,000
Buyer's total payment	$127,000

- **Cost plus award fee** (CPAF): a contract in which an award pool is created and managed by an award committee; subjective judgments are used to determine the award, giving the buyer more flexibility than in a CPIF contract; administrative costs are high

Time and Material Contracts

Time and material (T&M) contracts (sometimes called **unit price** contracts) contain characteristics of both FP and CR contracts and are generally used for small dollar amounts. These contracts may be priced on a per-hour or per-item basis (like a fixed price contract), but the total number of hours or items is not determined (open-ended cost type arrangements like a CR contract). A **purchase order** is a simple form of a unit price contract that is often used for buying commodities. It is a unilateral contract and only signed by one party, rather than a bilateral contract that is signed by both parties.

> **EXAM TIP**
>
> Projects with a high degree of "vagueness", such as a feasibility study, should use a time and materials contract type.

- An example of a T&M contract: in anticipation of a hurricane forecast for the site of a construction job, the buyer's project manager asks the contractor to prepare for high winds on a time and materials basis to be invoiced after the work is completed. The seller buys 20 8X4-foot plywood panels for a total of $1,000 and hires local labor for 15 hours at $10 per hour for a total of $150. The seller submits an invoice for $1,150.

Make-or-Buy Analysis

The make-or-buy analysis is a technique to determine if a particular work product should be purchased externally or produced internally within the performing organization. A make-or-buy decision is a product of this analysis.

Cost is usually a major factor in a **make-or-buy decision**. The actual or direct out-of-pocket costs to purchase a product as well as the indirect costs of managing procurement must be considered in any make-or-buy decision. Many other factors may also be considered in a make-or-buy decision. Some of the factors in this decision are:

EXAM TIP

Similar situations can yield very different make-or-buy decisions based on the factors influencing the project.

- Availability of human resources and skills
- Direct control/customization required
- Design secrecy and proprietary information
- Availability of product
- Reliability of suppliers
- Small volume requirements
- Limited capacity/time
- Production capacity
- Competition
- Degree of standardization
- Maintenance, support, and service required

Market Research

Organizations that employ various market research techniques to identify and qualify prospective sellers will be better prepared to define requirements and procurement objectives, ensuring that the most qualified and appropriate vendors are chosen.

Procurement Management Plan

The procurement management plan gives overall guidance for how procurements will be conducted, including types of contracts, risk response plans, standard documents to be used, how to work with the procurement department, defining schedule and lead time criteria, identifying pre-qualified sellers, etc.

Procurement Statement of Work

EXAM TIP

A procurement statement of work could be a source document for a subcontractor's project charter.

The procurement statement of work (SOW) is written to describe a procurement item in sufficient detail to allow prospective sellers to determine if they are capable of providing the item. Key details to know about a SOW are:

- The SOW describes the portion of the product to be purchased
- If the seller is producing the entire product, then the procurement SOW = the product description; in other cases, the product description is a broader definition of the project

- In government terms, the SOW describes a procurement item that is a clearly specified product or service, while a **statement of objectives** (SOO) is used for procuring an item that is presented as a problem to be solved
- Each procurement item needs its own SOW
- Multiple products and services may be grouped as one procurement item

Information in a SOW can include:
- Specifications on the product to be delivered
- The quantity desired to be produced
- The quality level desired
- Any performance data that must be adhered to
- The period of performance
- The work location

Procurement Documents

Procurement documents are another output of the Plan Procurement Management process. These documents are prepared by the **buyer** to help **sellers** understand the project's needs and to solicit proposals. Procurement documents provide information for prospective sellers that generally include:
- The **procurement statement of work**
- Background information for the product
- Procedures for replying
- Guidelines for preparation of the proposal
- Pricing forms
- A target budget or maximum
- Proposed terms and conditions of the contract
- **Evaluation criteria** for how proposals will be rated or scored

Procurement documents need to provide enough detail to ensure consistent, comparable responses from the sellers, yet allow potential sellers the flexibility to provide creative, viable solutions. Well-designed procurement documents enable:
- More complete proposal responses
- Facilitated comparisons of sellers' responses

12

- Pricing that is close to objectives
- Increased understanding of the buyer's needs and scope of work
- Decreased number of changes to project work

Procurement documents vary from industry to industry. The choice of which type of procurement document to use depends on the form of the **procurement statement of work** and the **contract type** selected. Common procurement documents include:

- **Invitation for bid** (IFB) or **request for bid** (RFB): the buyer requests a single price for the entire package of work; items of service are typically high dollar value and are standardized
- **Request for quotation** (RFQ): buyer requests a price quote per item, hour, etc.; items are relatively low dollar value; this may be used to develop information that may be put into a request for proposal

Source Selection Criteria

Evaluation criteria for how sellers will be selected are another output of the Plan Procurement Management process. Additional weight may be given to one or more criteria depending on the needs of the project. Evaluation criteria may include:

- Seller's price or cost criteria, as well as the overall life cycle costs (purchase plus operating costs)
- Seller's understanding of the project needs and requirements, seller's technical abilities, warranty, proprietary technology, and intellectual property rights
- Seller's past performance, management approach, ability to manage risk, and current working relationships
- Seller's financial capacity and reputation, capacity to produce the product, size, and type of business, including certification as a small, disadvantaged, or minority-owned enterprise

---

EXAM TIP

Evaluation criteria should be objective in nature and easily quantifiable.

12

CONDUCT PROCUREMENTS PROCESS

After the **procurement documents** have been sent out to prospective sellers, the Conduct Procurements process involves obtaining responses from these sellers. The key outputs of this process are **selected sellers** and the **agreement**.

The organization may conduct several types of competition among sellers:

- **Full, open competition**: all sellers are invited to bid; this may create more work for the department handling the bid process, and it may also lead to unqualified sellers being selected when there is a requirement to select the lowest bid
- **Pre-qualification of sellers**: sellers must qualify to bid by submitting information; a limited number of sellers will be selected to bid
- **Single source**: a seller is selected with no competitive bidding; this may be appropriate when there is an emergency, a specialized service is required, or the seller has entered into a long-term agreement with the buyer's organization; in government contracts, reasons for selecting a single source must be documented and approved
- **Sole source**: a seller is selected because that seller is the only provider of a product or service

Case Study Exercise

Exercise 12-1: There are several building supply vendors that can provide our materials, two local roof truss manufacturing companies, and several potential stucco sub-contractors we can choose from. Develop a purchase plan of the various vendors and determine 1) how we should solicit each vendor and 2) which type of contract would be best to use in each situation.

Procurement Negotiations

Negotiation is a general management skill that comes up frequently in project management. Although the primary objective of negotiation is to reach agreement on a fair and reasonable price, product, and delivery, it is important to look for a **win-win** situation in which a good relationship is developed with the seller. A **win-lose** situation may result in contract issues and problems for both the buyer and the seller.

Some of the key items to be considered in negotiation are:
- Responsibilities and roles
- Technical and business management approaches and methodologies
- Who has final authority
- Applicable laws
- Price
- Contract financing and payment terms

Negotiation Tactics

There are several negotiation tactics (adapted from PMI's *Principles of Project Management*):
- **Imposing a deadline**: a powerful tactic since it emphasizes the schedule constraints of the project and implies a possible loss to both parties
- **Surprises**: one party springs a surprise on the other, such as a change in dollar amount
- **Stalling**: one party claims it does not have the authority, that the person with authority is not available, or that it needs more information
- **Fair and reasonable**: one party claims the price is equitable because another organization is paying it
- **Delays**: necessary when arguments are going nowhere, tempers are short, or one party is off on a tangent
- **Deliberate confusion**: either distorting facts and figures or piling on unnecessary details to cloud the issues

- **Withdrawal**: either the negotiator is so frustrated that he or she does not continue or one side makes an attack and then retreats
- **Arbitration**: a third party is brought in to make decisions, including a decision on the final outcome
- **Fait accompli**: one party claims "What is done is done" and cannot be changed

Agreements

A procurement agreement is the major output of the Conduct Procurements process, which may be called a/an:
- Understanding
- Contract
- Subcontract
- Purchase order

A procurement agreement represents a formal agreement, and requirements must be stated in as much detail as possible. To manage agreements effectively, changes should be in writing and formally controlled by both the buyer and seller. A valid enforceable agreement allows disputes to be resolved in the court system or through alternative methods.

In the United States, commercial contracts are regulated by the Uniform Commercial Code (UCC); government contracts are regulated by the Federal Acquisition Regulation (FAR).

Black's Law Dictionary describes the ten tests for a contract. A project manager must ask whether a contract includes all of the following:
- Agreement between two or more parties
- Promise to do or not to do something
- An offer
- Unconditional acceptance
- Mutual assent: both parties agree to the terms
- Complete acceptance; there is no contract if there are modifications or partial acceptance

EXAM TIP
Situational questions on the exam will reflect appropriate conduct as it relates to United States' contract law.

- Consideration: something of value
- Legal capacity: of proper age and, within an organization, having the proper authority
- Legal purpose: does not violate public policy
- Form required by law: real estate contracts are required to be in writing in most jurisdictions

Organizations frequently have standard contracts that are preprinted. If signed as is, these are legally sufficient and will form a valid contract. Contracts are likely to include:

- A description of the **procurement statement of work**
- Terms for **inspection**, **warranty**, and support
- The **schedule**, where the work will be performed, and where products will be delivered
- The price, payment terms, fees, incentives, and penalties
- Any modifications to an organization's standard contracts
- How disputes will be resolved

When necessary, standard contracts may be modified to fit the requirements of a project. These are called **special provisions** and can be arranged by the project manager together with the **contract administrator** and legal counsel if necessary.

CONTROL PROCUREMENTS PROCESS

The Control Procurements process involves making sure that the seller is performing the work according to contractual requirements and that both the buyer and the seller are meeting their contractual obligations. Key project management processes that must be integrated with the Control Procurements process include:

- **Direct and Manage Project Work**: performing the work of the project plan
- **Control Quality**: inspect and verify non-conformance may lead to breach of contract

- **Perform Integrated Change Control**: ensure changes are properly managed and communicated; contested changes are called claims, disputes, or appeals
- **Control Risks**: review performance of the contracted work to make sure negative risks are reduced and that positive risks are enhanced

Contract Change Control System

Since project managers are often faced with multiple or ambiguous interpretations of a contract in deciding on the scope of work involved, there is high potential for conflict between the **buyer** and **seller**. Some buyer organizations may utilize a centralized group whose specific function is to handle contract administration. This group may contain one or more contracting officers or **contract administrators** who are assigned to various projects. In such cases, this person is the only one with authority to change a contract.

A **contract change control system** is a key tool and technique of the Control Procurements process. It defines the process by which the contract may be modified.

Contract change modifications fall into three categories:
- **Administrative change**: often a unilateral contract change, in writing, that does not affect the substantive rights of the parties
- **Change order**: a written order, signed by the contracting officer, directing the seller to make a change
- **Legal**: mutual agreement to modify a contract, documenting that all parties have considered and approved a change

Changes may be contested between **buyer** and **seller** when agreement cannot be reached. These may be called **claims**, **disputes**, or **appeals** and should be handled through a **claims administration** process. If changes are contested, they may be resolved in a court of law or through alternative **dispute resolution** procedures such as **mediation** or **arbitration**.

> **EXAM TIP**
> The contract administrator is typically a specialist in managing projects and ideally is not the project manager.

CLOSE PROCUREMENTS PROCESS

Each procurement item will be closed during the Close
Procurements process. This may occur at various times
during the project and will be part of the overall Close
Project or Phase process. The major objective of the Close
Procurements process is to verify that work has been
completed correctly and satisfactorily. It also involves
the administrative work of documenting the records and
lessons learned, resolving claims and disputes, and
completing any unique terms and conditions specified
in the contract.

Challenges in Closing Procurements

Some issues that could impact the Close Procurements
process are:

- **Waiver**: relinquishing rights under the contract;
  the **waiver pitfall** occurs when the project
  manager for the buyer knowingly accepts
  incomplete, defective, or late performance without
  objection; in doing so, the project manager waives
  his or her right to specific performance
- **Contract breach**: failure to perform a
  contractual obligation
- **Material breach of contract**: a significant
  failure to perform that may result in the non-
  faulted party being discharged from any further
  obligations under the contract
- **Time is of the essence**: when explicitly stated in
  the contract, the seller's failure to perform within
  an allotted time will constitute a material
  breach of the contract
- **Subcontract management**: if the seller
  subcontracts to a third party, the buyer has no
  direct control over the performance of the third
  party; the seller must manage all third parties
  and pay them

Effects of default for non-performance by the **seller** are:
- Seller is not entitled to compensation for work in process not yet accepted by the buyer
- Buyer is entitled to repayment from seller of any advance or progress payments applicable to such work
- Buyer may order delivery of completed or partially completed work
- Seller must preserve and protect property buyer has an interest in
- Seller is liable for excess procurement costs

Early termination of a contract is a special case of closing procurements. It may result from the:
- Mutual agreement of both parties
- Default of one party
- Convenience of the buyer

Closure Outputs

As organizational process assets are updated, the Close Procurements process will provide:
- **A procurement file**: a complete set of indexed records to be included with the final project archives when closing the phase or project; contract documents for documenting work and work performance
- **Deliverable acceptance**: formal written notice to the seller as defined in the contract
- **Lessons learned documentation**: documentation from **procurement audits** as well as an evaluation of the seller for future reference; recommendations for improving the procurement process

> **EXAM TIP**
> The PMI *Code of Ethics and Professional Conduct* requires fair dealing with all parties.

12

KEY INTERPERSONAL SKILLS FOR SUCCESS

The interpersonal skill highlighted in this chapter is:

Negotiation

Project managers need to develop negotiation skills because it's likely that some stakeholders will have competing interests and objectives for some aspects of any project. The goal is to reach an agreement that all parties can accept and support, even if they don't agree with every aspect of it.

The skills needed to successfully negotiate overlap with other interpersonal skills such as conflict management, communication, trust building, influencing, and leadership. Project managers need good analytical skills that will help them focus on the issues, distinguish between wants and critical needs, and discover what is most important to each party. Listening and not taking situations personally are vital. In addition, the *PMI Code of Ethics and Professional Conduct* requires fair dealing with all parties.

SAMPLE PMP EXAM QUESTIONS ON PROCUREMENT MANAGEMENT

1. Which one of the following statements about contract changes is correct?

   a) Contract changes seldom provide realizable benefits to the project
   b) A fixed price contract will minimize the need for change control
   c) Inspections or audits can be used to eliminate potential changes
   d) More detailed specifications will reduce the most common causes of changes

2. You are the project manager on a large government contract. You have determined that you will need to procure a major component of the project due to a lack of expertise within your existing project team. A technique you will most likely employ to find prospective sellers is:

   a) Inspection
   b) Advertising
   c) Expert judgement
   d) Make or buy analysis

3. A project manager was in the process of identifying which project deliverables could be met by procuring products and services from prospective sellers. This project manager was in the midst of which planning process?

   a) Plan Procurement Management
   b) Collect Requirements
   c) Resource Planning
   d) Conduct Procurements

4. A file that houses a complete set of contract documentation is called:

   a) Closed procurements
   b) Contracts file
   c) Lessons learned file
   d) Procurement file

5. What is the benefit of the process of obtaining seller responses and awarding a contract?

   a) Maintaining or increasing the effectiveness of stakeholder engagement
   b) Determining whether to acquire goods or services, when, and how much
   c) Ensuring that buyer and seller meet contract requirements
   d) Providing alignment of internal and external stakeholder expectations

6. Which of the following is part of the Control Procurements process?

   a) Answering the questions of potential sellers
   b) Developing a request for proposal
   c) Negotiating a contract
   d) Confirming that changes to a contract are made

7. A weighting system is used to do which of the following?

   a) Conduct a bidder conference
   b) Identify a sole source
   c) Develop independent estimates
   d) Rank proposals by score

Notes:

8. A contractor completes work as clearly specified within the SOW, but the buyer is not pleased with the results. In this case, the contract is considered to be:

   a) Complete because the contractor met the terms and conditions of the contract
   b) Incomplete because formal acceptance has not been provided by the buyer
   c) Incomplete because the specifications are incorrect
   d) Complete because the contractor is satisfied, and work results follow the SOW

9. You are the owner's representative looking for a reputable firm to build an apartment complex. You need a company that has experience in building apartment complexes within budget. Before distributing proposal requirements, you've asked the procurement department to come up with a list of locally-based firms that have a reputation for managing costs closely. This is an example of a:

   a) Contract negotiation
   b) Qualified sellers list
   c) Proposal evaluation technique
   d) Seller rating systems

10. Jacob was the project manager of an organization that entered into a cost plus fixed fee contract with XYZ Co. As the project progressed, the scope of the project increased with approved change orders. The cost of the project also increased. In such a case, the fee:

    a) Decreased in the same proportion as the cost changed
    b) Remained the same because it was a fixed fee
    c) Increased as the buyer absorbed the increased costs
    d) Was negotiated as part of the approved change orders

Notes:

12

11. You are the project manager on a large government contract. You procured a major component of the project due to a lack of expertise within your existing project team. Once the component has been delivered, you communicate to the contract administrator that the product is acceptable. The contract administrator will most likely:

    a) Cancel the contract because it's been completed
    b) Send an invoice to the seller
    c) Send the seller a written notice of acceptance
    d) Perform a procurement audit

12. One of the vendors you are procuring materials from in your project is behind schedule. He has promised that he will get caught up in the next two weeks but doesn't want you to report negatively back to the sponsor because it could jeopardize his ability for future business. He offers you season tickets for the local baseball team if you delay mentioning the concern until after the end of the two weeks. You should:

    a) Accept the offer and give the vendor the opportunity to rectify the situation
    b) Tell the vendor that they need to resolve the issue in 3 days, not 2 weeks
    c) Not accept the offer, but don't tell your project sponsors for at least two weeks
    d) Not accept the offer and report the truth to your project sponsor

Notes:

13. A contract has a clause that allows the buyer to cancel a contract for convenience. The buyer cancels the contract when about 30% of the work has been completed. What happens to the seller?

a) The seller will go through an alternative dispute resolution process
b) The seller is likely to start litigation to get the buyer to complete the contract
c) The seller must receive payment from the buyer for work completed
d) The seller will cancel the remaining portion of the contract for default

14. Which of the following best describes what is included in procurement agreements?

a) The buyer will manage a project and the seller will make sure the contract items meet the needs of the project
b) Procurement agreeements show that the terms understanding, contract, subcontract, and purchase order have the same meaning
c) They include terms, conditions and specifics about how the seller must perform
d) It is the seller's responsibility to assure that agreements follow organizational procurement policies

15. The Plan Procurements Management process identifies:

a) Which project needs can be best met by buying products
b) What sellers propose as a solution to the request
c) How payments will be scheduled for the contract
d) What contract change control will be used

Notes:

ANSWERS AND REFERENCES FOR SAMPLE PMP EXAM QUESTIONS ON PROCUREMENT MANAGEMENT

Section numbers refer to the *PMBOK® Guide.*

1. **D Section 12.3 – Monitoring and Controlling**
   B) is part of the Conduct Procurements process; C) is part of the Control Procurements process; D) is not generally part of procurement management; policies and procedures should become part of organizational process assets, after approval.

2. **B Section 12.2 – Executing**
   Many government contracts require that major procurements are advertised to as many prospective sellers as possible.

3. **A Section 12.1 – Planning**
   The purpose of the Plan Procurement Management process is to identify which project needs can be best met by purchasing products, services, or results.

4. **D Section 12.4 – Closing**
   The procurement file also includes the closed contract.

5. **D Section 12.2 – Executing**
   A) is conducted in the Manage Stakeholder Engagement process; B) is conducted in the Plan Procurement Management process; C) is conducted in the Control Procurements process.

6. **D Section 12.3 – Monitoring and Controlling**
   A) occurs in the Conduct Procurements process; B) occurs in the Plan Procurement Management process; C) occurs in the Conduct Procurements process.

7. **D** **Section 12.2 – Executing**
   A) is a technique to answer bidders questions in an open forum; B) a sole source is the only provider of a product or service and would not need to be ranked; C) the purpose of independent estimates is to have a benchmark to compare to sellers' proposals.

8. **B** **Section 12.4 – Closing**
   Although the contract work is complete, the contract is not closed. The buyer must provide written notice to the contractor that the contract has been completed. Without this formal acceptance and closure, the contract cannot be closed. However, the contractor can demand that the buyer close out the project because work was done to the agreed-on specifications of the SOW.

9. **B** **Section 12.1.1.9 – Planning**
   A) occurs after you have selected the seller during the Conduct Procurements process; C) and D) are different types of criteria used to select sellers.

10. **D** **Section 12.3 – Monitoring and Controlling**
    A) the fee as a percentage of the total contract may decrease, but the dollar amount of the fee will not decrease; B) it's unlikely that the seller will agree to do more work for the same fee; C) having the buyer absorb the cost will not affect the fee in a cost reimbursable contract.

11. **C** **Section 12.4 – Closing**
    A) canceling the contract is different from closing the contract; B) the seller will send an invoice based on payment schedules that have been structured in the contract; D) procurement audits are a tool and technique of the Close Procurements process.

12

**12. D Section 5.3.1 – Monitoring and Controlling
(*Code of Ethics and Professional Conduct*)**
As a project manager, your integrity and ethics are
of utmost importance. A) accepting bribes is not
ethical; B) is not the best answer as it forces an
unrealistic constraint on the vendor; C) is not the
best answer since you should not hide facts about
a project.

13. C Section 12.4 – Closing
A) and B) if termination for convenience is allowed,
it's unlikely the seller would have any basis for
dispute resolution or litigation; D) there is no
indication that anyone defaulted on the contract.

14. C Section 12.2 – Executing

15. A Section 12.1 – Planning
B) is part of the Conduct Procurements process;
C) and D) are part of the Control Procurements
process.

CASE STUDY SUGGESTED SOLUTION

Exercise 12-1:

Vendor list for the Lawrence Garage Project

General Building Materials Vendors

These are commonly available commodities and should be solicited via an invitation for bid. We will supply the bill of material (BOM) from the architect. Selection will be based solely on price. Once selected, we will send the winner a purchase order for the materials listed in the BOM and set up a delivery schedule based on our plan.

Truss Manufacturers

Trusses are essentially custom made for each building, so we should solicit via a request for proposal (RFP). Because the proposed solutions may be significantly different, we will have to evaluate each for its impact on our budget and schedule. Once we decide, a firm fixed price (FFP) contract should be issued since we will know exactly how many trusses are needed.

Stucco Contractors

We expect the stucco subcontractors to supply all their own materials, including tarpaper, rigid foam insulation, chicken wire, and the ingredients that make up the stucco. We should solicit via an RFP for all work and materials. Because the prices of some of the materials used are going up rapidly, the contract should be a combination of cost reimbursable for the materials and firm fixed price for the balance.

STAKEHOLDERS

CHAPTER 13 | **STAKEHOLDERS**

STAKEHOLDER MANAGEMENT

Project success may be determined more by meeting the needs and expectations of stakeholders than by any other factor. While project teams may focus on bringing in the scope of the project, with the appropriate quality, on time, and on budget, the knowledge area of stakeholder engagement has taken on a new importance in managing projects successfully. This knowledge area has a general focus on stakeholders outside the project team.

Stakeholder management doesn't just include communicating better with stakeholders. As you deepen your understand of these concepts, you will begin to see the complexities that underlie the analysis of stakeholders as well as the types of interrelationships and competing interests that must be resolved for projects to deliver what is desired.

Stakeholder identification has become more important because even the smallest projects often have an impact on a wide range of stakeholders, and those stakeholders may have an unexpected effect on a project. Part of identification includes determining a stakeholder's attitudes about the project, the product of the project, the team, and management. Add to that mix a stakeholder's ability to influence the project, either formally or informally, and this situation can create potent opportunities and threats for project managers.

Analyzing stakeholders to plan how you should engage them in a project is likely to require discretion on the part of the project manager and team. Implementing the stakeholder management plan and controlling stakeholder engagement will include a variety of communication strategies and interpersonal skills. Issues related to stakeholders will challenge even the most experienced project managers.

> **EXAM TIP**
> For the PMP exam, it is especially important to understand the needs of stakeholders who are not part of the project team.

13

Things to Know

1. The four processes in stakeholder management:
   • **Identify Stakeholders**
   • **Plan Stakeholder Management**
   • **Manage Stakeholder Engagement**
   • **Control Stakeholder Engagement**
2. Stakeholder analysis techniques to **identify stakeholders** and their **expectations**
3. Components of a **stakeholder analysis**
4. The **power/interest grid** and **salience model**
5. The **stakeholder register**
6. How to conduct a **stakeholder engagement assessment**
7. The **stakeholder management plan**
8. **Techniques for managing stakeholder expectations**
9. Key **interpersonal skills** for success

Key Definitions

Affiliation power: power that results from whom you know or whom an individual has access to.

Change log: a comprehensive list of changes made during a project.

Expert power: power that results from an individual's knowledge, skills, and experience.

Legitimate power: formal authority that an individual holds as a result of his or her position.

Penalty (coercive, punishment) power: power that results from an ability to take away something of value to another.

Referent (charisma) power: power that results from a project manager's personal characteristics.

Reward power: power that results from an ability to give something of value to another.

Stakeholder: an individual, group, or organization who may affect, be affected by, or perceive itself to be affected by a decision, activity, or outcome of a project.

Stakeholder register: a project document that includes the identification, assessment, and classification of project stakeholders.

Virtual team: a group of people with a shared goal who fulfill their roles with little or no time spent meeting face to face.

IDENTIFY STAKEHOLDERS PROCESS

Because project managers want to meet stakeholder needs and expectations, it's important to determine who the stakeholders are and what interests they have in a project. Stakeholders may be actively involved, they may be affected by a project, or they may have influence over a project.

There are several parts to the Identify Stakeholders process: identifying stakeholders, assessing expectations and requirements, analyzing stakeholders' attributes, and classifying stakeholders. Inputs to this process include the project charter, which may define internal and external stakeholders, procurement documents, enterprise environmental factors, and organizational process assets.

Identifying Stakeholders

Stakeholders may be found from expected sources, such as project team members, functional departments, end users, customers, sellers, contractors, consultants, internal or external specialists, regulators, the public, etc. Project managers should consider how operational stakeholders may affect or be affected by a project. Some such stakeholders may include sales and plant managers, customer call center employees, manufacturing operators, and maintenance workers.

> **EXAM TIP**
> Stakeholders can have varied and also conflicting needs for communication.

> **EXAM TIP**
> An important technique in identifying stakeholders is expert judgement. Project managers must look outside of their own experience to those who may have had experience with stakeholders of the current project.

13

Stakeholder Expectations

The project manager will try to develop as much detail as possible about stakeholder needs, how those needs translate to requirements for the product of the project, how the defined scope will meet those needs, and expectations for their own involvement. Stakeholders often have competing needs and interests, and these needs and interests may vary during the project life cycle.

Stakeholder Analysis

EXAM TIP

Every project initiated must have an expected business value that all stakeholders on the project team must understand and agree to.

Once stakeholders have been identified, the project manager will conduct an analysis of their interests, influence, and involvement. The goal is to develop an approach to managing stakeholders and an appropriate way to communicate. The project manager may assess, classify, and prioritize stakeholders based on their involvement at different stages in the project life cycle.

These assessments of stakeholders are in terms of their:
- Power: legitimate authority or other types of power over the project, control of resources, and source of control over the project
- Interest and attitude: a stakeholder's beneficial or detrimental stake in the project, support, opposition, or indifference to the project
- Influence: level of participation, contribution, and connection to the project
- Impact: the effect of a stakeholder's influence
- Relationships: which stakeholder's cooperate or are in conflict
- Risk profile: the stakeholder's risk attitudes, including risk appetite, risk tolerance, and risk thresholds; this process intersects with the Plan Risk Management process

13

Power/Interest Grid

Once assessments have been conducted, models may be used to combine the attributes evaluated. A power/interest grid like the one shown in Figure 13-1 below combines these two attributes and helps teams identify stakeholders who will support the project, ignore the project, and potentially detract from the project.

	Low Interest	High Interest
High Power	High Power Low Interest	High Power High Interest
Low Power	Low Power Low Interest	Low Power High Interest

Figure 13-1
Power/Interest Grid

Case Study Exercise

Exercise 13-1: List at least 5 key stakeholders you would consider important in the Lawrence Garage Project. Identify each stakeholder's level of power and interest.

Salience Model

The salience model, developed first by Mitchell, Agle, and Wood in the late 1990s, measures the intersection of power, urgency (the stakeholder's time requirements), and legitimacy (a combination of a stakeholder's authority and appropriateness of involvement). See the example in Figure 13-2 on the following page.

13

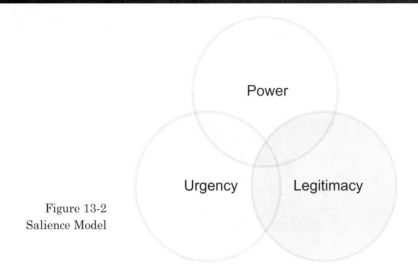

Figure 13-2
Salience Model

Stakeholder Register

One goal throughout these assessments is to determine stakeholders' responses to potential project situations. All of the identification and analysis work is documented in the stakeholder register. This register will include demographic information, assessment information, and classifications of stakeholders.

PLAN STAKEHOLDER MANAGEMENT PROCESS

Following the identification and analysis of stakeholders, the project manager needs to decide how to deal with stakeholders. The plan should document action steps to engage stakeholders, manage their expectations, and maintain relationships among team members and other stakeholders.

The **stakeholder register** and **project management plan** are main inputs to this process, along with enterprise environmental factors and organizational process assets. From the project management plan, the defined scope, human resource requirements, and communication techniques will have an impact.

Stakeholder Engagement Assessment

One major goal of this process is to determine the level of stakeholder engagement the project manager needs to ensure during the project life cycle. Just as project team members may have differing levels of work during the project, so stakeholders may need to have different levels of involvement throughout the life of a project. A stakeholder engagement assessment matrix is one tool for defining the current state (C) and desired state (D). See Figure 13-3 below for an example.

Stakeholder	Role	Unaware	Resistant	Neutral	Supportive	Leading
Jackie K	Sponsor				D C	
Henry W	PM					D C
Kat R	Designer		C			D
Maria G	Implementer		C		D	
Lou B	Implementer			C	D	

Figure 13-3
Stakeholder
Engagement
Assessment

Engagement levels will vary from stakeholder to stakeholder and from project to project. Engagement levels can range from unaware (not connected to the project and not understanding the project's potential impacts), to leading (totally aware of the project and its potential impacts and actively engaged in the project). Defining stakeholders' engagement levels can significantly improve communication and ultimately project success.

Stakeholder Management Plan

Outputs to the Plan Stakeholder Management process are project document updates and a **stakeholder management plan**. The stakeholder management plan includes:

- Engagement levels (both current and desired)
- Impact (effects of scope changes on stakeholders)
- Relationships and overlap among stakeholders
- Communication requirements
- Information on distribution format and content
- Need for communication and reason
- Timing of information distribution

> **EXAM TIP**
> Current and desired states are "perceived" states.

MANAGE STAKEHOLDER ENGAGEMENT PROCESS

In the Manage Stakeholder Engagement process, the goal of the project manager is to increase support for project efforts and to minimize any opposition stakeholders may have. The project manager needs strong management skills in keeping others on track, identifying and resolving issues, making progress toward the objectives, and anticipating situations related to stakeholders that may derail the project. Project managers also need strong interpersonal skills in building trust, negotiation, influence, communication, listening, conflict management, and maintaining stakeholders' interest and commitment.

> **EXAM TIP**
> The stakeholder management plan should be a "living" document and used and reviewed regularly to ensure stakeholder needs and interests are being consistently addressed throughout the project life cycle.

This process relies on the **communications management plan** and the **stakeholder management plan** to guide the actions project managers will take regarding stakeholders. Since changes are likely to occur during a project's execution, other inputs include the **change log** and **organizational process assets**.

13

Techniques for Managing Stakeholder Expectations

There are many techniques project managers can use to manage stakeholder expectations, but at a minimum, project managers should:

- Influence expectations to increase the likelihood of project acceptance
- Address concerns and anticipate future problems
- Resolve issues or risks that have occurred, which may result in project changes

To accomplish these objectives, the communication methods the project manager can use include frequent and regular meetings, developing and tracking issues via an **issues log**, and resolving conflict. Face-to-face meetings provide an optimal means of communicating. When a stakeholder is virtual, other means such as teleconferencing, web meetings, phone calls, instant messaging, email, or other electronic tools must be used extensively to ensure effective communication. For stakeholders with high power, project managers should consider using primarily one-on-one communication.

The outputs of this process include updates to project documents and the project management plan. There are many updates that may be made to organizational process assets, along with the issues log and lessons learned documentation.

Case Study Exercise

Exercise 13-2: Using your list of key stakeholders from Case Study Exercise 13-1, create a stakeholder register that defines each stakeholder's role and their current and desired states.

> **EXAM TIP**
>
> The project manager has the responsibility to ensure the stakeholders and the project team are in alignment. You can do this by reviewing the project charter with the stakeholders and obtain approval on key deliverables, milestones, and roles and responsibilities.

EXAM TIP
By holding regular stakeholder meetings, or by providing appropriate communication, the project manager can manage the flow of information and keep stakeholders engaged and informed.

CONTROL STAKEHOLDER ENGAGEMENT PROCESS

The Control Stakeholder Engagement process focuses on maintaining stakeholder engagement throughout the project by monitoring relationships and revising strategies for managing stakeholders when needed.

The **issues log**, **work performance data**, and project documents are inputs to this process that help the project manager determine what is happening with stakeholders. In addition, some of the specific elements in the project management plan will be inputs to the Control Stakeholder Engagement process. These inputs include the plans to accomplish work, managing all the core project team members, supporting team members and others who are doing the work of the project, and the **change management process**.

As the project manager uses the stakeholder management plan to monitor stakeholder activities and engagement, he or she will also link this plan to the scope of work completed, verification of quality, and validation of scope. Any discrepancies may be identified for corrective or preventive actions.

13

KEY INTERPERSONAL SKILLS FOR SUCCESS

The interpersonal skills highlighted in this chapter are:

Political Awareness

Project managers must learn how to navigate organizational politics, which are always present. Politics in an organization are established by norms of behavior, communication, how power is used, etc. Politics affect how the project management processes are carried out, and project managers need to find appropriate ways to work within the system and, potentially, when to go outside the system.

The basic assumptions driving business changed during the 1990s. Before, what was discussed were things like continuity, planning, diversification, scale, security, uninformed customers, and national borders as driving business factors. Now what is discussed are change, coping with the unexpected, focus, segmentation, flexibility, responsiveness, speed, demanding customers, and freedom of movement as the most prevalent factors driving business.

Culture

Culture is an important concept in stakeholder management. In today's society in which the global economy is influencing more and more organizations, it is important for a project manager to understand how culture can impact a project and what can be done to ensure success in a multicultural project environment.

Culture is everything that people have, think, and do as members of their society and is shared by at least one other person. It is important to understand that culture is learned; it is not bred into a particular group. Culture is a cluster of related values that individuals have in common and believe in. Can people's beliefs be influenced? Of course they can. What an individual believes in his or her childhood is influenced by experiences and encounters throughout his or her life, which will shape that person's value system.

> **EXAM TIP**
> Not all projects go smoothly. The project manager may have to use influencing skills in order to keep stakeholders aligned and communication flowing effectively.

> **EXAM TIP**
> All cultures experience continual cultural change.

13

Culture is very complex yet dynamically stable. In other words, culture evolves over time. Cultural changes do not happen overnight; they are adaptations to human learning, environmental conditions, and the interactions between people's learning and conditions.

Cultures are integrated from the perspective that a particular value may be related to other values within a culture; cultures are not random assortments of values.

Cultural differences require different management approaches by the project manager. Knowing the values of each project participant will greatly improve the overall effectiveness of the project manager. Cultural differences should also be considered a value-add to the project. Diversity provides an opportunity for new solutions to old problems. Do not shy away from a culturally diverse project. You may learn a lot!

Trust Building

Trust is the foundation of effective teamwork. Without trust, it's difficult for teams to be honest and share their ideas and opinions. Without trust, it will be more difficult for teams to make decisions, hold each other accountable, and measure progress and accomplishment accurately. Project managers must make the effort to build trust among core team members as well as key project stakeholders and influencers.

SAMPLE PMP EXAM QUESTIONS ON STAKEHOLDER MANAGEMENT

1. Jason, the project manager, was instructed to meet the deadline for a website update project "no matter what" because of a marketing campaign for a new product. He had identified the stakeholders and sent them copies of the schedule when their part of the project work was due. In spite of this, they were late with their deliverables, and the team had to work extensive overtime to make the project deadline. What should the project manager do to reduce the likelihood of this happening on the next project?

   a) Create a stakeholder management plan for engagement
   b) Classify stakeholders by their level of interest and influence
   c) Monitor stakeholder relationships and adjust the strategies for dealing with them
   d) Foster better relationships between stakeholders and the project team

2. The primary purpose of identifying stakeholders is to:

   a) Get their help in developing the WBS
   b) Create a project organization chart
   c) Update the organizational process assets
   d) Analyze their expectations of the project

3. You are the project manager of the implementation of a new product line in a manufacturing facility in southeast Asia. The project is almost complete. The project has passed the quality control inspections and has met the quality requirements. All documented issues have been addressed, and many of the resources have been released. The project is slightly ahead of schedule but has a small budget overrun. The sponsor has called a meeting to get final signoff when you receive an email from the customer, asking for a major change. As the project manager, what should you do?

a) Discuss the change with the project team to determine what it would take to implement
b) Get more detailed information from the customer on the change and what is needed
c) Suggest to the customer that the change should be handled in the next project phase
d) Inform the sponsor of the customer's request and ask for the signoff to be delayed

4. Sara identified a number of concerns that an information technology database manager had about the timing of when resources would be needed on Sara's project because of an unrelated implementation project. When Sara prepared the schedule, there was no problem. Now the information technology database manager's implementation project has been extended due to unexpected problems, and the resource may not be available for Sara's project. What should Sara do first?

a) Manage stakeholder expectations through negotiation
b) Identify and assess the risks associated with not having the database resource on time
c) Clarify and resolve any other issues that have been identified
d) Conduct a stakeholder analysis of the information technology manager's interest and involvement

Notes:

5. You are the project manager starting a new project, and many of the stakeholders identified in the charter are the senior executives of the current organization, as is the customer. One way to determine a strategy for managing communications with these stakeholders would be to:

   a) Perform a power/interest analysis
   b) Interview the stakeholders together to gain consensus on their expectations
   c) Create a stakeholder register
   d) Perform a probability and impact analysis

6. Which of the following processes provide an input to the Manage Stakeholder Engagement process?

   a) Direct and Manage Project Work
   b) Perform Integrated Change Control
   c) Control Communications
   d) Manage Project Team

7. As the project manager of a very complex business process change project, you have several groups of stakeholders who will be negatively impacted by the project delivery. It is your responsibility to engage these groups and ensure a successful implementation. You employ many different methods to overcome the expected resistance to change and are able to gain the trust and confidence of these impacted groups. This is an example of using _____ skills.

   a) Negotiating
   b) Public speaking
   c) Management
   d) Interpersonal

Notes:

13

8. Nikhil is the project manager for an information technology upgrade to a customer relationship management system. There are several new features in the software that the sales department is especially interested in using to their fullest capabilities. These features require integration between sales, order entry, and shipping. As the project work is completed, what should Nikhil review to make sure the sales department will be satisfied?

   a) Stakeholder register
   b) Risk register
   c) Work performance data
   d) Stakeholder management plan

9. Emily, the project manager, identified stakeholders and discovered that the director of marketing and controller were not in agreement. What should she do first?

   a) Classify stakeholders by their level of interest and influence
   b) Seek expertise to help sort out the needs and expectations of the two stakeholders
   c) Monitor stakeholder relationships and adjust the strategy for dealing with them
   d) Foster better relationships with stakeholders and the project team

10. Which of the following organizational process assets are most likely to affect the Identify Stakeholders process?

   a) Organizational culture and regional practices
   b) Stakeholder analysis and industry standards
   c) Templates for the stakeholder register and lessons learned
   d) Product life cycle and quality policies

Notes:

13

11. As outputs of the Manage Stakeholder Engagement process, all of the following may be updated except for the:

a) Issues log
b) Stakeholder register
c) Communications management plan
d) Enterprise environmental factors

12. To assess how involved a stakeholder might be in a project, you should use which of the following tools?

a) Responsibility assignment matrix
b) Stakeholder register
c) Stakeholders engagement assessment matrix
d) Salience model

13. Why should a project manager spend time identifying and classifying stakeholders early in a project?

a) To assure ease of approvals from the regulators
b) To understand a stakeholder's ability to influence the project
c) To monitor stakeholder engagements efficiently
d) To minimize resistance to the project

14. You are assigned as a new project manager of a project that is 40% complete. You review the current issues logs and meeting minutes. You determine that there are three critical issues that have not yet been resolved, and these issues appear to be the reason for the stakeholders' frustration. One of the first actions you must take is to:

a) Review the outstanding issues with the stakeholders and discuss options for corrective action
b) Update the communications management plan
c) Increase the frequency of communication to stakeholders
d) Perform a contingency reserve analysis to determine the amount of cost reserves available

Notes:

13

15. Which of the following are outputs of the Control Stakeholder Engagement process?

    a) Organizational process assets updates and meetings
    b) Work performance information and change requests
    c) Issue logs and project management plan
    d) Project documents and expert judgment

Notes:

ANSWERS AND REFERENCES FOR SAMPLE PMP EXAM QUESTIONS ON STAKEHOLDER MANAGEMENT

Section numbers refer to the *PMBOK® Guide*.

1. **A Section 13.2 – Planning**
   It sounds like the project manager stopped after identifying stakeholders; B) is a technique used in the Identify Stakeholders process; C) these activities take place in the Control Stakeholder Engagement process; D) are the actions taken in the Manage Stakeholder Engagement process.

2. **D Section 13.1 – Initiating**
   Understanding stakeholders' expectations of the project as well of their level of interest in the project will assist the project manager in planning the project appropriately.

3. **B Appendix X3.6 – Executing**
   A) trying to solve the problem before you understand it is futile, so make sure you have time to evaluate the change; C) pushing the customer to the next phase is inappropriate until you have examined the nature of the change; D) evaluate the change before postponing the meeting.

4. **B Section 13.3 – Executing**
   Choices A and C are not necessarily the case in all projects; D) is not a bad answer, but the question doesn't necessarily define the size and agenda items of the presentation; B) is the most correct answer because the development of the presentation is added work that will need to be performed on the project to support the stakeholder's needs.

5. **A Section 13.1 – Initiating**
   A power/interest analysis (as part of stakeholder engagement) can identify the potential impact or support each stakeholder can generate.

6. **B** **Section 13.3.1 – Executing**
A change log is an input to the Manage Stakeholder Engagement process; the change log is an output of the Perform Integrated Change Control process.

7. **D** **Section 13.3 – Executing**
Overcoming resistance to change is an interpersonal skill that is critical in this kind of situation.

8. **C** **Section 13.4.1 – Monitoring & Controlling**
All of these may be reviewed, but as the project is being executed, work performance data are an input to the Control Stakeholder Engagement process.

9. **B** **Section 13.2.2.1 – Planning**
A) should have taken place when the stakeholders were identified; C) take place in the Control Stakeholder Engagement process; D) are the actions taken in the Manage Stakeholder Engagement process.

10. **C** **Section 13.3.1.4 – Executing**
A) are enterprise environmental factors; B) stakeholder analysis is a tool of the Identify Stakeholders process, not an input; industry standards are an enterprise environmental factor; D) product life cycle and quality policies are an organizational process asset, but they are not as useful as answer C.

11. **D** **Section 13.3.3 – Executing**
Enterprise environmental factors are neither inputs nor outputs to the Manage Stakeholder Engagement process.

13

12. **C** **Section 13.2.2.3 – Planning**
A) is used to document assignment of responsibilities to team members; B) is used to document stakeholder identification; D) is a technique to identify a stakeholder's power, urgency, and legitimacy.

13. B Section 13.1 – Initiating
A) it may help understand what regulators need, but this would take place in the Plan Stakeholder Management process; C) is part of the Control Stakeholder Engagement process; D) is part of the Manage Stakeholder Engagement process.

14. A Section 13.3 – Executing
It is always best to take the time to understand the issues before acting. In B), there is no need to update the communications management plan if there is no need to change; in C), there is no need to increase the frequency of communication unless the cause of the issues was a lack of communication; D) does not apply because a reserve analysis compares the amount of the contingency reserves remaining to the amount of risk remaining.

15. B Section 13.4.3 – Monitoring & Controlling
A) an organizational process asset is an output and meetings are techniques in the Control Stakeholder Engagement process; C) are both inputs to the Control Stakeholder Engagement process; D) expert judgment is a technique in the Control Stakeholder Engagement process.

13

CASE STUDY SUGGESTED SOLUTIONS

Exercise 13-1

Stakeholder	Power	Interest	Comments
Mr. Lawrence	High	High	
Mrs. Lawrence	High	Medium	She wants her husband to be happy
Scott Hiyamoto (Architect)	Medium	High	
Scott Hiyamoto (Project Manager)	High	High	
Neighbors	Low	Medium	Mr. Lawrence has already talked to his neighbors on either side of his house
Acme Construction & Engineering [Your name here]	Medium	High	
Susan Ruzicka (General Manager, ACE)	Medium	Medium	

Exercise 13-2

Stakeholder	Role	Unaware	Resistant	Neutral	Supportive	Leader
Mr. Lawrence	Customer					C D
Mrs. Lawrence	Customer			C		D
Scott Hiyamoto	Architect				C D	
Scott Hiyamoto	Project Manager					C D
Neighbors			C	D		
ACE [Your name here]	Contractor				C D	
Susan Ruzicka	General Manager			D	C	

FINAL EXAM

SAMPLE FINAL EXAM

If you take this sample final exam, you should spend 65 minutes answering these 50 sample questions. This timing is similar to the average time per question used in the actual PMP exam.

1. A project has a 60% chance of finishing on time and a 30% chance of finishing over budget. What is the probability that the project will finish on time and within budget?

    a) 12%
    b) 18%
    c) 42%
    d) 48%

2. Given that BAC = 100, EV = 20, PV = 40, and AC = 50, what is the TCPI for the project if the project team must achieve the BAC?

    a) 0.875
    b) 1.14
    c) 1.60
    d) 2.00

3. A project had 7 team members when 3 new team members joined the team. How many additional communication channels will there be?

    a) 21
    b) 24
    c) 42
    d) 45

Notes:

4. You have awarded a cost-plus-incentive fee (CPIF) contract for your project. The target cost is $300,000, and the target fee is $24,000. The fee share ratio is 80-20, with a minimum fee set at $10,000. What would be your total payout on the contract if the seller realized an actual cost of $400,000?

   a) $400,000
   b) $404,000
   c) $410,000
   d) $424,000

5. You have been volunteering for your local PMI chapter and have agreed to present a 4-hour mid-day session on an advanced topic next Saturday. Then a friend gives you tickets for next Saturday's after-noon game of your local baseball team against a team you've been wanting to see. You've tried to find a replacement to give your talk, but it's too short notice. What should you do?

   a) Ask PMI to postpone the session
   b) Give half the session and finish early
   c) Cancel the session
   d) Go to the game late and miss the first 4 innings

6. The customer has asked for major changes that will incur significant costs. If the project is near completion, the project manager should:

   a) Ask the customer for a description of and reasom for the change
   b) Say no to the customer because the project is almost complete and there is less cost flexibility
   c) Meet with the project sponsor and determine if the changes should be made
   d) Meet with the project team to determine who can work overtime

Notes:

14

7. A member of your team brings ideas for enhancements to the project product to a team meeting. These suggestions will add work to the project that is beyond the requirements of the project charter. As project manager, you point out that only the work required for the project should be completed by the team. Any request for enhancements should be formally submitted through the:

a) Change control system
b) Change control board
c) Control Scope process
d) Configuration management board

8. During which process will a project manager address how to satisfy the completion of work, the transfer of the project's product into operations, and the collection of records?

a) Close Procurements
b) Direct and Manage Project Work
c) Close Project or Phase
d) Monitor and Control Project Work

9. You have been working hard to resolve conflict on a project team that resulted from unclear scope requirements. What is a likely output of this effort?

a) Colocation
b) Change requests
c) Team building activities
d) Training

Notes:

14

10. You are the project manager for a medical device
company. You hear that one of the quality
assurance analysts falsified some of the test results
based on encouragement from the director of
marketing. You are fairly certain the information
is true. You should:

   a) Confront the quality assurance analyst privately
   to confirm the allegations
   b) Confront the quality assurance analyst in the
   next team meeting
   c) Contact the director of marketing and ask if he
   or she has instructed the quality assurance
   analyst to falsify records
   d) Talk to your sponsor and determine a best course
   of action to validate the accusations

11. William has identified risks on a project. What
should he do with this information?

   a) Consider the effects on the cost estimate and
   quality plan
   b) Revise the network diagram and project schedule
   c) Review seller responses to an RFP and select a
   seller
   d) Direct the project team to assign responsibilities
   to the stakeholders

12. Due to the departure of the prior project manager,
you are assigned to a project that is 40% complete.
You review the current issues logs and meeting
minutes, and you determine that there are a few
project team members that have not been delivering
work in a timely manner. The prior project manager
had not provided any feedback to their manager on
their performance. One of your first actions might be:

   a) Contact the team members and review their
   deliverables and assess the reason for task delays
   b) Discuss the issue in the weekly project status
   meeting and gain consensus of the entire team
   c) Develop recommended corrective actions
   d) Add 4 weeks to the project schedule as a
   contingency

Notes:

14

13. You are planning a project to update the facilities in a medical office. Cost and schedule are both important in this project. What should be included in your schedule management plan to help control the schedule?

    a) Cost estimates for the work
    b) Work breakdown structure
    c) Activity duration estimates
    d) Schedule control thresholds

14. As the project manager for a business process improvement project for a strategic business process, you have been given an estimate of $3 million to deliver a project within 12 months. The project is a key strategic initiative this year. If the project is delivered, the organization will experience a 30% reduction in process costs. In developing the budget, one action that should result from your review of the estimate is likely to be:

    a) Create a risk register of cost risks and planned responses
    b) Aggregate the work package estimates to create a budget baseline
    c) Submit a change request for the additional resources
    d) Perform a bottom-up estimating exercise

15. A project manager new to project management has been given an unapproved project charter, and he comes to you for advice. What should you tell him to do first?

    a) Create a work breakdown structure
    b) Start documenting the project constraints and assumptions
    c) Find out who should approve the project charter and obtain approval
    d) Begin obtaining resources for his team

Notes:

14

16. As the project manager on a project which is similar to one previously managed, you decide to utilize a _____ WBS in order to facilitate activity definition with the team.

    a) Detailed
    b) Template
    c) Organizational
    d) Planning

17. You are underway on an important project when a key resource, John, just mentioned to you that he will be going on vacation for 3 weeks in the middle of the project. His manager has identified a less experienced resource as a backup. One of the first things you should do is:

    a) Determine if John knows and approves of the resource
    b) Schedule a team building exercise to facilitate his or her entry to the project team
    c) Identify if the new resource's skill level will impact the durations of the schedule activities
    d) Double the estimate for the activities the new resource will be picking up for John

18. You have taken over a project to update your corporate website. It is about 30% complete, and you know there is a project management plan available. What should you do first?

    a) Develop a WBS
    b) Define the statement of work
    c) Identify risks
    d) Collect work performance data

Notes:

14

19. As the project manager on a medical device project, you are asked to put together a budget for the project. You put together a budget of $3 million, but about one month into your project, new information is presented that proves that your budget was $300,000 too low due to work that was part of the project scope but unidentified in the planning. As the project manager, you should:

   a) Do nothing
   b) Prepare a revised EAC using the CPI method
   c) Obtain the money from the management reserves
   d) Report the variance to the customer and ask for more money

20. A communication skill that project managers should develop to clarify and confirm understanding is:

   a) Facilitating
   b) Listening
   c) Presentation
   d) Meeting management

21. It seems that projects in your organization never fully close. You want to avoid this on your current project. What could you do to close your current project?

   a) Finalize all procurement payments and documents
   b) Make sure the scope goes through the validation process
   c) Conduct a quality and procurement audit
   d) Provide step-by-step actions to ensure work has been completed

22. Which process will assist the project manager in minimizing resistance to a project?

   a) Identify Stakeholders
   b) Plan Stakeholder Management
   c) Manage Stakeholder Engagement
   d) Control Stakeholder Engagement

Notes:

14

23. You are on a project team for a grocery chain that is creating marketing materials for an upcoming promotion to tie in with national heart health month. You have learned that your internal print department is overwhelmed with work and will not be available to complete the point of sale advertising materials. What should you do first?

    a) Recommend that the company put off the heart health promotion until next year.
    b) Document the situation in the issue log and communicate to the marketing department.
    c) Find an external printer and get them working on the marketing materials immediately.
    d) Ask your sponsor to help your project get a higher priority with the internal print department.

24. Michael has reviewed the project charter, identified stakeholders, and begun development of the project plan. He is looking at the existing risk templates, categories, and lessons learned. What should he do next?

    a) Perform a Monte Carlo simulation to determine an appropriate schedule.
    b) Tailor the definitions of probability and impact to his project.
    c) Develop a risk adjusted cost estimate based on past performance.
    d) Use the risk scoring guidelines to develop a high-level assessment of overall risk.

25. Your project is nearing completion, and you have scheduled a deliverable review meeting with your customer for next week. The objective of the meeting will be to verify that each project deliverable has been completed satisfactorily and has been accepted by the customer. This is an example of the _____ process.

    a) Management by Objectives
    b) Validate Scope
    c) Control Scope Process
    d) Configuration Management

Notes:

26. One of the major problems facing project managers in acquiring staff is that functional managers are not willing to give up their best people. In some organizational structures, when functional managers provides resources, they do not give up control over salary and promotion of their people. This is a disadvantage of the _____ organizational structure.

a) Matrix
b) Projectized
c) Virtual
d) Life cycle

27. A meeting to answer all the potential sellers' questions is called a:

a) Negotiated settlement
b) Procurement statement of work review
c) Bidder's conference
d) Proposal evaluation meeting

28. A salience model is a combination of which stakeholder characteristics?

a) Power, interest, and involvement
b) Influence, impact, and interest
c) Power, influence, and impact
d) Urgency, power, and legitimacy

29. You are the contract administrator of an apartment complex construction project. Quality Construction, Inc. is the builder that you've hired. You decide to conduct a structured review of quality audits conducted during the seller's work. This is an example of:

a) A procurement performance review
b) A contract type
c) Inspections
d) Claims administration

Notes:

14

30. The project leader who works as a staff assistant and communications coordinator, but who cannot personally make decisions is called a:

    a) Project coordinator
    b) Project expediter
    c) Matrix manager
    d) Project manager

31. Portfolio, program, project, and organizational project management all differ in how each contributes to achieving an organization's strategic goals. Which one harmonizes project and program components to realized specific benefits?

    a) Project management
    b) Portfolio management
    c) Organizational project management
    d) Program management

32. A primary output of the Close Project or Phase process is:

    a) The project management plan
    b) Final result transition
    c) Accepted deliverables
    d) Contract documentation

33. To document how specific requirements meet the business need of a project, the team would use:

    a) A project management plan
    b) Requirements documentation
    c) A requirements traceability matrix
    d) A project scope statement

Notes:

14

34. You are the project manager for a business process improvement project for a strategic business process. Sue and Bob work at a different facility than the rest of the project team and they bring expertise in the critical business process being redesigned. In order for them to work successfully, it is imperative that the project manager address this virtual team in the:

   a) Project schedule
   b) Time management plan
   c) Communication management plan
   d) Scope management plan

35. You are the project manager starting a new project, and many of the stakeholders identified in the charter are the customer or the senior executives of the current organization. After performing a stakeholder analysis, you've identified that most of the stakeholders fall into the "Monitor" quadrant. However, the President of the customer company is categorized as "Manage Closely." With this information you should:

   a) Regularly check in with the President to ensure that he is completing his tasks on time
   b) Plan for a regular written status report for the President of the customer company and a weekly face-to-face meeting with the rest of the project team
   c) Include all the stakeholders in the weekly team meeting to make sure that all the stakeholders are provided all the necessary information they need
   d) Plan for a regular written status report for most of the stakeholders and a weekly face-to-face meeting with the President of the customer company

Notes:

14

36. As project manager, you notice that a team member is not performing well because he is inexperienced in the technology being utilized. What is your best solution?

    a) Report the team member's bad performance to his functional manager
    b) Colocate the team member with someone more experienced with that technology
    c) Discuss a reward mechanism with the team member to encourage him to work harder
    d) Arrange for the team member to get training in the required technology

37. Maria has received work performance data and project documents that indicate one of the stakeholders has not signed off on a major deliverable. She discovers that the stakeholder is not receiving information in a timely manner. What should she do next?

    a) Review the issues log to determine if there are critical factors that were missed
    b) Bring the team in for a face-to-face meeting to make sure everyone gets the information
    c) Find a way to use the information management system more effectively
    d) Use the expertise of an external consultant to review the business process

38. The degree, amount, or volume of risk that an organization or individual will withstand is called a:

    a) Risk attitude
    b) Risk appetite
    c) Risk tolerance
    d) Risk threshold

Notes:

39. You have a complete WBS and activity list and are preparing to put the list in sequence for your project to exhibit at a trade show. One of the work packages is to install the trade show booth. What will help you define the flow of activities for this deliverable?

   a) Schedule management plan
   b) Project scope statement
   c) Milestone list
   d) Network diagram

40. You have been assigned to an internal project to install a new inventory system and need to put together a quality management plan. This project is similar to one you've worked on before, and the customer has agreed to provide ten resources to help test the new inventory system. You decide to:

   a) Build a new quality plan based specifically on the requirements of the new inventory system
   b) Utilize the quality management plan from the last project, as is, since it was successful
   c) Let the quality assurance manager put together a new quality management plan
   d) Implement the prior plan, but keep only what's relevant to the new inventory system

41. The main output of the Identify Stakeholders process includes which items?

   a) Stakeholder demographics and classification
   b) Assessment information and a communication plan
   c) Stakeholder management plan and an action plan
   d) Identification information and a stakeholder management plan

Notes:

14

42. Identifying the influence a stakeholder has on a project is typically recorded in the:

    a) Stakeholder register
    b) Risk register
    c) Stakeholder management strategy document
    d) Project charter

43. During the negotiation process, once an agreement is reached, you should do all of the following EXCEPT:

    a) Stay positive and respectful
    b) Define the agreement as specifically as possible
    c) Continue negotiating
    d) Express your desire for a long and prosperous relationship

44. You are a project manager in a fairly immature project environment, and your customer has not provided you with a charter. In order for you to create a project scope statement, you should:

    a) Stop work until the charter is received
    b) Ask the sponsor to step in and talk to the customer
    c) Send a team to the customer location to complete the scope statement
    d) Obtain an informal written description from the customer

Notes:

14

45. Use the following chart for the next question:
Which task is behind schedule and under budget?

Task	PV	AC	EV
1	95	100	95
2	150	130	110
3	130	130	130
4	80	60	70

   a) Task 1
   b) Task 2
   c) Task 3
   d) Task 4

46. You have just been notified that your customer has money problems and will not be able to pay for the upcoming milestone deliverables. As project manager, you should:

   a) Tell everyone to stop working
   b) Release 90% of the project team
   c) Reduce the scope and begin administrative closure
   d) Shift these deliverables to the next phase to give the customer additional time to obtain funds

47. A key stakeholder on your construction project has been spending time at the construction site, interrupting the team's efforts at trying to stay on schedule. As the project manager, you should:

   a) Tell the team not to talk to the stakeholder and send him to you
   b) Do nothing because the stakeholder can do whatever he wants
   c) Talk to the stakeholder and communicate how his interest in seeing how the job is progressing is actually hampering the team's efforts
   d) Talk to the stakeholder and tell him that he cannot come out to the construction site anymore

Notes:

14

48. Reserve analysis compares:

   a) The effectiveness of the risk management process to the project objectives
   b) Current project cost estimates with similar past project costs
   c) Project performance to the planned schedule
   d) The amount of contingency reserves remaining to the amount of risk remaining

49. What is a potential problem if activity durations are estimated before team members are assigned?

   a) The team may not operate according to ethical standard for fairness
   b) The activity durations may change based on the competency of the team member
   c) The project manager will need to influence the end user to accept a change in scope
   d) Quality assurance activities may need to be revised to accommodate benchmarking

50. Henry is the project manager for a business process improvement project for a strategic business process that is of key importance to revenue generation. The key stakeholders have requested a weekly written report on the project. Henry suggests that a central website be created for all stakeholders and key measures be posted to that website regularly. This is an example of:

   a) Work performance data
   b) Managing communications
   c) A communications management plan
   d) Information management systems

Notes:

SAMPLE FINAL EXAM ANSWERS with explanations and references can be found in Chapter 15, Appendix B.

APPENDICES

APPENDIX A

SAMPLE ASSESSMENT EXAM ANSWERS AND REFERENCES

Section numbers refer to the *PMBOK® Guide.*

1. B **Section 13.3 – Stakeholder Management – Executing**
    A) a lessons learned knowledge base is a historical record of the outcomes of previous projects; C) a quality audit addresses project policies and procedures; D) A formal project review is not necessary; the best answer B) is designed to start increasing communication with the stakeholders.

2. B **Section 10.1.2.3 – Communication Management – Planning**
    You must know the sender-receiver model of communication in order to answer many questions on the exam.

3. A **Section 8.2.2.3 – Quality Management – Executing**
    A) a process analysis may include a root cause analysis; B) quality audits are used to audit the project policies, processes, and procedures, NOT the results of a project; C) a defect repair review is a verification that a defect has been repaired based on the defined deliverables of the project; D) a Pareto diagram identifies which defects are more common than others.

4. D **Section 10.3 – Communications Management – Monitoring and Controlling**
    A breakdown in communication implies there is a lack of information, misunderstanding, conflict among stakeholders, or some other negative situation. The most correct answer is "usually negatively."

5. B **Section 6.2.3.2 – Time Management – Planning**
    Activities have duration and are used for detailed project planning and scheduling; milestones have no duration and may be at a high level or detailed.

15

6. A **Section 7.4.2.1 – Cost Management – Monitoring and Controlling**
   There are several ways to calculate EAC.
   B) if you assume costs will continue at the budgeted amount, then EAC = AC + (BAC – EV)
   EAC = \$5,625,000 + (\$12,600,000 – \$6,300,000) = \$11,925,000
   C) to calculate actual costs, begin with the basic formula:
   CPI = EV ÷ AC
   Use algebra to revise the formula to:
   AC = EV ÷ CPI
   AC = \$6,300,000 ÷ 1.12 = \$5,625,000
   AC = EV ÷ CPI, so CPI = EV ÷ AC
   D) If you assume cost performance will continue at the same rate, then EAC = BAC ÷ cumulative CPI
   EAC = \$12,600,000 ÷ 1.12 = \$11,250,000

7. D **Section 10.2.2.5 – Communications Management – Executing**
   Performance reporting is part of Manage Communications.

8. D **Section 5.4.3.1 – Scope Management – Planning**
   The project, work packages, or activities must be defined before estimating; A) analogous estimating relies on parameters from previous projects; B) bottom-up estimating requires that scope has been decomposed to work packages or activities; C) discovery is not a *PMBOK® Guide* term.

9. A **Section 6.5.2.2 – Time Management – Planning**
   B) bottom-up estimating is the most detailed estimating technique; C) three-point estimating is not as quick because you would obtain three estimates for each activity; D) resource leveling is not focused on cost estimating; it's a technique to adjust the schedule based on resource constraints.

10. A **Section 13.1 – Stakeholder Management – Initiating**
    All objectives of the project should be met; therefore, managing customer expectations is NOT the best answer, nor is choosing between objectives.

11. C   Section 11.5 – Risk Management – Planning

A promotion would be the trigger to force the project team to execute the risk response plan.

12. C   Section 12.4 – Procurement Management – Closing

A procurement audit is used to identify successes and failures that warrant recognition in the preparation or administration of other procurement contracts.

13. C   Section 5.2.3.2 – Scope Management – Planning

The requirements traceability matrix is an input to the Control Scope process. A) the matrix may be used to communicate, but it does not necessarily control communications; for B) and D), the matrix is focused on linking requirements and deliverables, not stakeholders or a schedule.

14. D   Domain III – Task 6 – Human Resources Management – Executing

Although all of these suggestions may work, establishing ground rules and guidelines for acceptable behavior by team members would be the most effective.

15. C   Section 4.5 – Integration Management – Monitoring and Controlling

A) contingency reserves address the additional funds or time needed to minimize the risk of overruns; B) the Control Procurements process is the process of managing the contract and the relationship between the buyer and seller; a certain amount of configuration management will be used by the contract administrator; D) the Perform Integrated Change Control process includes reviewing, approving, and managing changes; configuration management is one part of this process.

16. B   Section 12.1 – Procurement Management – Planning

All of the other types of contracts listed are cost reimbursable contracts; risks that cause changes in cost will be reimbursed; profit is a negotiated amount, plus any negotiated incentives.

17. A   Section 11.6 – Risk Management – Monitoring and Controlling

B) a threat is something negative that would affect the schedule by making it late; a positive schedule variance is likely to be due to an opportunity that will bring in the project early; C) a negative cost variance means the project is over budget; D) an SPI greater than 1.0 means you are ahead of schedule; if it cannot be explained, you probably need to investigate.

18. B   Section 9.2.2.2 – Human Resources Management – Executing

Negotiating for critical resources is a key technique that project managers should use to improve the probability of project success.

19. D   Section 6.3.2.2 – Time Management – Planning

A) discretionary dependencies are those that are not required, but are based on some level of experience or best practice; B) external dependencies are those that are typically outside the project team's control; C) contractual dependencies are also mandatory but they are not inherent in the nature of the work; an inherent dependency would occur in a home construction project where the walls must be built before the roof can be put on.

20. A   Section 10.2.3 – Communications Management – Executing

Lessons learned are outputs of the Manage Communications process and should be shared with all stakeholders; B) lessons learned should be included in a project knowledge database with access by all authorized individuals; C) in addition, just sharing the results with the team is not optimal either; D) lessons learned reviews should not be blaming sessions; you must take great care in making these reviews a learning experience so that people will openly participate in future reviews.

21. C   Section 4.5 – Integration Management – Monitoring and Controlling

C) is a slightly better answer than B); most likely there will be no question as to the urgency of this legislative change, and a vote won't necessarily change the need for incorporating the change.

22. A   Section 8.0 – Quality Management – Planning
During project planning, it is important to understand the expected levels of grade and quality that are needed; these levels will help in determining the level of quality control that will be required on the project.

23. D   Section 11.4 – Risk Management – Planning
The party should be held outside because the expected monetary value is $1,600, which is greater than the $920 to be earned if it is held inside. Of course, you can only choose whether to hold the party inside or outside; you cannot choose whether it rains or not.

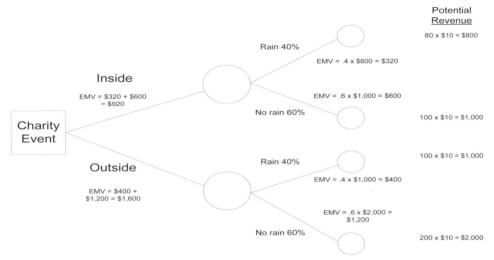

24. A   Section 13.1 – Stakeholder Management – Initiating
Key benefits for the following are: B) you will define a plan to interact with stakeholders; C) this helps gain support for the project and reduce opposition; D) during this process, you adjust your strategy for dealing with stakeholders, which will make your interactions more effective.

25. A   Section 12.4 – Procurement Management – Closing
Although the contract work is complete, the contract is not closed. The buyer must provide written notice to the contractor that the contract has been completed. Without this formal acceptance and closure, the contract cannot be closed. However, the contractor can demand that the buyer close out the project as work was done to the agreed-upon specifications of the SOW.

15

26. C Section 13.2 – Stakeholder Management – Planning
Although some of these answers appear to be correct, before embarking on a campaign to demonstrate your team's effectiveness, the best thing you can do is to LISTEN to what the new stakeholder will want from you. Engage in active listening and review the stakeholder analysis.

27. A Section 10.2 – Communications Management – Executing
B) is part of the Plan Communications Management process; C) may be true, but it is not the best answer; D) is part of the Control Communications process.

28. A Section 11.0 – Risk Management – Planning
B) unknown risks cannot be identified; C) a risk register documents known risks; D) responses are determined after risks have been identified and analyzed.

29. A Section 13.3 – Stakeholder Management – Executing
Clarifying and resolving issues that have been identified is a key part of the Manage Stakeholder Engagement process, along with addressing concerns that have not yet become issues and activity managing stakeholder engagement.

30. A Section 5.5 – Scope Management – Monitoring and Controlling
B) is WBS creation; C) takes place in the Control Quality process; D) takes place in the Control Scope process.

31. D Section 4.5 – Scope Management – Monitoring and Controlling
A) it is typically not the responsibility of the project manager to act as the change control board; B) the product requirements and project charter already exist; C) you are not accepting any completed deliverables.

32. B Section 4.3.1.3 Integration Management – Executing
A), C), and D) are all organizational process assets.

33. A Section 1.4.4 – Integration Management – Initiating
B), C), and D) all have the roles of project managers and PMOs reversed.

34. C **Section 4.1.3 – Integration Management – Initiating**
A) the project scope statement defines the project deliverables and the work required to create those deliverables; B) the project SOW is an input to the charter; D) a contract can define the SOW, which in turn is an input to the project charter and subsequently the scope statement.

35. A **Section 2.1.3 – Human Resources Management – Closing**
A projectized organization derives its revenue from performing projects. Once a project is complete, there may not be another project to move to.

36. B **Section 4.6 – Integration Management – Closing**
This is a special case that often requires additional documentation and input from stakeholders.

37. B **Section 13.4.2.3 – Stakeholder Management – Monitoring and Controlling**
A) an issues log is an input to the Control Stakeholder Engagement process, but it doesn't directly help the project team resolve disagreements; B), C), and D) are all tools and techniques of the Control Stakeholder Engagement process, but major disagreements are usually resolved best with face-to-face discussions.

38. A **Section 7.2.1 – Cost Management – Planning**
B) are outputs; C) and D) are tools and techniques of the Estimate Costs process.

39. B **Code of Conduct 1.5 – Human Resources Management – Initiating**
A), C), and D) are mandatory (required) standards; B) is an aspirational standard.

40. A **Section 7.3.3.1 – Cost Management – Planning**
Contingency should be planned for early in order to accommodate uncertainty in the project budget.

15

41. A Section 5.6 – Scope Management – Monitoring and Controlling
Before these new designs become a cost issue, the project manager must facilitate the discussion to determine if these requested changes will be in or out of scope.

42. A Code of Conduct 2.3.1 – Cost Management – Executing
The cleanest way to handle the whole thing is to not include the customs fee because it may have the appearance of a bribe.

43. B Section 9.4.2.2 – Human Resources Management – Executing
Even an enthusiastic team member can cause problems on a project; as initial tasks are completed, begin to understand how each of the team members perform, and respond individually if anyone impacts the plan.

44. D Section 9.3.2.3 – Human Resources Management – Executing
When there is a change in the team, or some major aspect of the project, it's likely the team will go through some of the team development phases again; A) the team is not adjourning because it still has significant work to complete; B) is norming, but the team was storming; C) in high performing teams, members want to share and hear other's opinions, but this team was not collaborating.

45. C Section 1.3 – Integration Management – Monitoring and Controlling
Project constraints are interdependent, so a change in one will almost always affect another.

46. C Section 8.1.2.3 – Quality Management – Monitoring and Controlling
Scatter diagrams show relationships between two variables; the closer the points are together, the stronger the relationship.

47. D Section 11.6.2 – Risk Management – Monitoring and Controlling

A project manager must be aware of the objectives of the project and must respond appropriately when environmental situations change.

48. B Code of Conduct 5.3.1 – Procurement Management – Monitoring and Controlling

The *PMI Code of Ethics and Professional Conduct* advocates that the project manager be honest and is responsible for seeking out and understanding the truth. While there is a risk the schedule may slip, the contractor said earlier it would be on time. The better answer is to report you are obtaining an updated status.

49. A Section 7.1 – Cost Management – Planning

B) the cost management plan wouldn't be updated until you review the assumptions used to develop the budget and thoroughly understand the budget rationale; C) and D) are not acceptable.

50. D Section 12.2 – Procurement Management – Executing

A) cost may be an important criterion, but this question doesn't specify that it is the only criteria; B) a bidder's conference takes place before proposals are submitted; C) the project manager may participate in the selection process, but D) is a better answer.

15

APPENDIX B

SAMPLE FINAL EXAM ANSWERS AND REFERENCES

Section numbers refer to the *PMBOK® Guide*.

1. **C** **Section 11.4.2 – Risk Management – Planning**
   On time = 60%; within budget = 100% – 30% (over budget)
   60% X 70% = 42%

2. **C** **Section 7.4.2.3 – Cost Management – Monitoring and Controlling**
   There are 2 methods to calculate TCPI. If it must be based on the BAC, then the formula is:
   TCPI = (BAC – EV) ÷ (BAC – AC)
   TCPI = (100 – 20) ÷ (100 – 50) = 80 ÷ 50 = 1.6

3. **B** **Section 10.1.2.1 – Communications Management – Planning**
   Total number of channels = (n X (n – 1)) ÷ 2
   10 team members = (10 X 9) ÷ 2 = 45
   7 team members = (7 X 6) ÷ 2 = 21
   Additional = 24

4. **C** **Section 12.3 – Procurement Management – Monitoring and Controlling**
   Comparing the target cost ($300,000) to the actual cost ($400,000) shows a $100,000 overrun. This overrun is shared 80-20 by the buyer and seller (the seller's share is the one shown last). In this situation, 20 percent of the $100,000 overrun is $20,000, which is deducted from the target fee of $24,000, resulting in an adjusted fee of $4,000. However, the contract sets the minimum fee at $10,000. As such, the seller receives a fee of $10,000 plus the actual costs incurred of $400,000, for a total payout of $410,000.

5. **D** **Code of Conduct 4.3.2 – Integration Management – Executing**
   According to the code of ethics, you have a duty of loyalty to PMI because you made a commitment to speak before you were offered the tickets.

6. A **Section 5.6 – Scope Management – Monitoring and Controlling**
The project manager must first determine if the requested changes impact the business need the project was undertaken to address; if the project justification, product, deliverables, or objectives are impacted, the request is valid and the project sponsor must be notified.

7. B **Section 4.5 – Integration Management – Monitoring and Controlling**
Although A) is a good answer, the submission of a request should be provided to the change control board; B) is a better answer.

8. C **Section 4.6 – Integration Management – Closing**
These are all outputs of the Close Project or Phase process. A) the Close Procurements process may involve the completion of work, but not the full completion of the project and the other closing activities; B) outputs of the Direct and Manage Project Work process are deliverables, work performance data, and other documentation about project progress and updates; D) outputs of the Monitor and Control Project Work process focuses on change requests, work performance information, and other documentation.

9. B **Section 9.4.3 – Human Resource Management – Executing**
A) Colocation is useful for improving communication and helping people work better together, but it doesn't specifically relate to scope; B) change requests are an output of the Manage Project Teams process; C) and D) are techniques of the Develop Project Team process and are useful to help resolve conflict.

10. D **Section 2.3.1 – Quality Management – Executing**
As the project manager, you have an obligation to report any unethical or illegal behavior; you may need to strategize with your sponsor before confronting anyone.

11. A **Section 11.2 – Risk Management – Planning**
Be sure to review the data flow diagrams for each knowledge area process.

12. A Chapter 9.4.2.2 – Human Resources Management – Executing
The best answer is to talk to the people to determine the exact situation before reacting to the issues.

13. D Section 6.1.3.1 – Time Management – Planning
The goal is to know when schedule variances exceed the threshold and are unacceptable.

14. B Section 7.3.3.1 – Cost Management – Planning
A) creating a risk register is not an output of the Determine Budget process, but updating it MAY be; B) is the best answer because, once the budget is finalized, a cost baseline should be established; C) there isn't anything to warrant a change request at this time; D) a bottom-up estimate was hopefully performed prior to you receiving the budget.

15. C Section 4.1 – Integration Management – Initiating
The project charter provides the authority to proceed. Without the authority, the project manager faces future problems.

16. B Section 6.2.1.4 – Time Management – Planning
Template WBSs may be detailed or not, but will usually include standard activity lists that are applicable to your organization's particular area.

17. C Section 6.5 – Time Management – Planning
The best answer is to assess the competency of the resource before making any changes to the schedule; A) although it would be nice to get John's input, it is not required; B) is nice to do, but is not the first thing you should consider; D) without proper assessment, just doubling an estimate is not the best approach.

18. D Section 4.3 – Integration Management – Executing
A), B), and C) are all planning activities, not executing. Since the project management plan exists, the planning activities related to the WBS, statement of work, and risk should be complete and you should review work performance data.

19. C **Section 7.4.3 – Cost Management – Monitoring & Controlling**
A) the variance here is only 10%, so it may not be necessary to do anything if that is acceptable to the project sponsor and customer; B) you may prepare a revised EAC, but it's not likely that the CPI method would be appropriate because this is an atypical event; C) reserves are used for unidentified work that is part of the project scope, which is exactly the situation here; D) you are not required to report the variance until after your evaluation is complete.

20. B **Section 10.2 – Communications Management – Executing**
A) brainstorming, conflict resolution, problem solving, and meeting management are facilitation techniques; C) abilities to tailor the material, maintain energy and interest, and engage the audience are key presentation skills; D) the ability to balance the need for thorough discussion without getting off topic is a key meeting management skill.

21. D **Section 4.6 – Integration Management – Closing**
A) is good practice for external purchases; B) is good practice but does not address other aspects of the project; C) a quality audit is part of a control process, not a closing process.

22. C **Section 13.3 – Stakeholder Management – Executing**
A) identifying stakeholders is necessary for all of the rest of the stakeholder management activities, but nothing in this process minimizes resistance; it helps you understand stakeholders' perspectives through the analysis that you do; B) it's important to plan how you will deal with resistant stakeholders based on your analysis of their interests, power, etc.; D) monitoring stakeholder relationships and communications is important to assure that you are dealing with them effectively.

**23. B Section 10.3 – Communications Management –
Monitoring and Controlling**
All of these may ultimately be part of the solution. The first
step is to document what has happened and make sure the
departments involved know about it. They may have other
information that will help you develop a solution.

24. B Section 11.1.3 – Risk Management – Planning
A) occurs after risks have 1) been identified, 2) prioritized
(qualitative assessment) and 3) quantified; C) comes after
either qualitative or quantitative analysis, which is after
planning and identification (P/I); D) comes after you have 1)
defined P/I guidelines, 2) identified risks, and 3) used the P/I
matrix in your assessment.

**25. B Section 5.5 – Scope Management – Monitoring and
Controlling**
A) management by objectives is a technique for focusing on
outcomes and can be useful in general project management;
B) inspection is a tool in the Validate Scope process that
includes activities such as measuring, examining, and
verifying; C) the main activity in the Control Scope process
is monitoring progress and managing changes; D)
configuration management is part of the Perform Integrated
Change Control process and focuses on managing changes to
the product.

**26. A Section 2.1.3 – Human Resources Management –
Executing**
Project managers typically have no direct authority over sal-
ary and promotions in a matrix organization.

**27. C Section 12.2.2.1 – Procurement Management –
Executing**
Bidder conferences are sometimes called contractor
conferences, vendor conferences, or pre-bid conferences.

28. D Section 13.1.2.1 – Stakeholder Management – Initiating
There are many types of models to evaluate stakeholders' attitudes toward a project. The salience model combines their ability to influence the project (power) with their need for attention (urgency) and the appropriateness of their involvement (legitimacy).

29. A Section 12.3.2.2 – Procurement Management – Monitoring and Controlling
B) contract types are associated with the financial aspect of a contract, not the review aspect; C) inspections are used to verify compliance in the seller's work processes or deliverables; D) claims administration occurs when buyers and sellers have differences about changes in the project.

30. B Section 2.1.3 – Human Resources Management – Initiating
Know the types of organizational structures and the responsibility levels for the project coordinator, expediter, and manager. A) the project coordinator has some authority to make decisions and report to a higher level than a project expediter; C) matrix manager is not a term used in the *PMBOK® Guide*; D) the project manager's authority varies according to the project's organizational structure.

31. D Section 1.4 – Integration Management – Initiating
Know the differences between project, program, portfolio, and organizational project management.

32. B Section 4.6.1 – Integration Management – Closing
A), C), and D) are all inputs to the Close Project or Phase process.

33. B Section 5.2.3.1 – Scope Management – Planning
In this case, the more specific answer is better. A) requirements documentation would be included in the project management plan; C) this matrix links project requirements to deliverables; D) this statement describes project scope, deliverables, major assumptions, and constraints.

34. C Section 9.2.2.4 – Human Resources Management – Executing
A virtual team is a benefit in that it creates new possibilities for team acquisition; however, it adds additional challenges to successfully communicating within the team.

35. D Section 13.2 – Stakeholder Management – Planning
Since the President is identified as "Manage Closely," the project manager will need to make an extra effort to keep that stakeholder fully informed.

36. D Section 9.3.2.2 – Human Resources Management – Executing
While B) is acceptable, D) is the best answer because the project manager's responsibilities include obtaining project-specific training for the development of team members.

37. C Section 13.4.2.1 – Stakeholder Management – Monitoring and Controlling
A) an issues log is an input to the Control Stakeholder Engagement process; B) may be what is needed in the short term, but C) is a better long-term solution; D) it may be better to look for an internal solution before bringing in a consultant.

38. C Section 11.0 – Risk Management – Planning
A) risk attitude is the overall perspective on the level of risk that an individual or organization is willing to take; B) risk appetite is the degree of uncertainty offset by the anticipation of a reward; D) risk threshold is the dividing line between accepting or not accepting risk.

39. B Section 6.3.1.5 – Time Management – Planning
In this case, the work defined in the scope in terms of the characteristics of the display will influence the flow of activities, e.g. do you need to design the banners, or do you have them in stock?

40. D Section 8.1 – Quality Management – Planning
A) there is no need to develop a completely new plan if an organizational process asset is available; B) existing organizational process assets should be tailored to the current project; C) is possible, but not the best answer because, even if the quality management plan is prepared by someone else, the project manager is responsible for ensuring that it is appropriate.

41. A Section 13.1 – Stakeholder Management – Initiating
B) assessment information is an output of the Identify Stakeholders process, but a communication plan is the output of the Plan Communications Management process; C) and D) the stakeholder management plan is an output of the Plan Stakeholder Management process.

42. A Section 13.1 – Stakeholder Management – Initiating
The stakeholder register is the primary output of the Identify Stakeholders process. It is used as an input to develop the strategy as part of Plan Stakeholder Management.

43. C Section 12.2 – Procurement Management – Executing
You should stop negotiating when an agreement is reached.

44. D Section 5.3.1.2 – Scope Management – Planning
It is always better to get something in writing from the customer as a starting point; A), B), and C) are things you may consider if you are unable to get an initial response from a customer.

45. D Section 7.4.2.1 – Cost Management – Monitoring and Controlling
For behind schedule, look for a task with a negative SV and an SPI that is less than 1.0.
$SV = EV - PV = 70 - 80 = -10$
$SPI = EV \div PV = 70 \div 80 = .875$
For under budget, look for a task with a positive CV and a CPI that is greater than 1.0.
$CV = EV - AC = 70 - 60 = +10$
$CPI = EV \div AC = 70 \div 60 = 1.16$

46. C Code of Conduct 2.2.4 – Integration – Closing
The best choice is to give the customer value for the money already spent and close out the project; it is your responsibility to advise stakeholders who need to know.

47. C Section 13.3 – Stakeholder Management – Executing
The best answer is C) because it has the project manager working with the stakeholder to explain the situation in a calm and objective way. Hopefully, the stakeholder in this situation will understand and try to minimize, if not eliminate, the interruptions.

15

48. D **Section 11.6.2.5 – Risk Management – Monitoring and Controlling**

As the project progresses, the project team will use schedule and cost reserves for risks that become issues. Reserve analysis is a look at what is left to cover risks that may affect the remainder of the project.

49. B **Section 9.0 – Human Resources Management – Planning**

Activity durations are estimates of calendar time needed to complete project activities. That time may be affected by the availability of specific people resources, the ability to apply multiple people to a task, and the skill and experience of each specific person. Activity duration estimates should take these into consideration for the specific person doing the work, which is why you should have that person involved in developing the estimates.

50. B **Section 10.2 – Communications Management – Executing**

A) work performance data is an input to the Control Communications process; C) a communications management plan is an output of the Plan Communications Management process; D) are a tool and technique of the Control Communications process.

APPENDIX C

GLOSSARY

This glossary is a supplement to the *PMBOK® Guide*. Some of these terms are not in the *PMBOK® Guide*'s glossary but may be used in test questions.

100% Rule: the WBS should represent the total work at the lowest levels and should roll up to the higher levels so that nothing is left out, and no extra work is planned to be performed.

Accuracy: the assessment of correctness.

Acknowledge: indicates receipt of a message by a receiver, but does not indicate that the receiver understood or agreed.

Active listening: the receiver confirms listening by nodding, eye contact, and asking questions for clarification.

Activity attributes: similar to a WBS dictionary because they describe the detailed attributes of each activity. Examples of these attributes are description, predecessor and successor activities, and the person responsible for completing an activity.

Activity contingency reserve: budget for a specific WBS activity within the cost baseline that is allocated for identified risks that are accepted and for which contingent or mitigating responses are developed.

Affiliation power: power that results from whom you know or whom an individual has access to.

Application area: a category of projects that share components that may not be present in other categories of projects. For example, approaches to information technology projects are different from those for residential development projects, so each is a different application area.

Authority: the right to make decisions necessary for the project or the right to expend resources.

Backward pass: used to determine the late start and late finish dates of activities.

Benchmark: comparing actual or planned project practices to those of comparable projects to identify best practices and generate ideas for improvement. Benchmarks provide a basis for measuring performance.

Bidder conference: the buyer and potential sellers meet prior to the contract award to answer questions and clarify requirements; the intent is for all sellers to have equal access to the same information.

Business value: the entire value of the business; the total sum of all tangible and intangible elements.

Buyer: the performing organization, client, customer, contractor, purchaser, or requester seeking to acquire goods and services from an external entity (the seller). The buyer becomes the customer and key stakeholder.

Capability maturity model integration (CMMI): defines the essential elements of effective processes. It is a model that can be used to set process improvement goals and provide guidance for quality processes.

Change control: the procedures used to identify, document, approve (or reject), and control changes to the project baselines.

Change log: a comprehensive list of changes made during a project.

Change management: the process for managing change in the project. A change management plan should be incorporated into the project management plan.

Chart of accounts: the financial numbering system used to monitor project costs by category. It is usually related to an organization's general ledger.

Code of accounts: the numbering system for providing unique identifiers for all items in the WBS. It is hierarchical and can go to multiple levels, each lower level containing a more detailed description of a project deliverable. The WBS contains clusters of elements that are child items related to a single parent element; for example, parent item 1.1 contains child items 1.1.1, 1.1.2, and 1.1.3.

15

Colocation: project team members are physically located close to one another in order to improve communication, working relations, and productivity.

Commercial-off-the-shelf (COTS): a product or service that is readily available from many sources; selection of a seller is primarily driven by price.

Constraints: a restriction or limitation that may force a certain course of action or inaction.

Contingency plan: a planned response to a risk event that will be implemented only if the risk event occurs.

Contingency reserve: budget within the cost baseline or performance measurement baseline that is allocated for identified risks that are accepted and for which contingent or mitigating responses are developed.

Contract: the binding agreement between the buyer and the seller.

Control account: the management control point at which integration of scope, budget, and schedule takes place and at which performance is measured.

Crashing: using alternative strategies for completing project activities (such as using outside resources) for the least additional cost. Crashing should be performed on tasks on the critical path. Crashing the critical path may result in additional or new critical paths.

Crashing costs: costs incurred as additional expenses above the normal estimates to speed up an activity.

Critical path: the path with the longest duration within the project. It is sometimes defined as the path with the least float (usually zero float). The delay of a task on the critical path will delay the completion of the project.

Decision theory: a technique for assisting in reaching decisions under uncertainty and risk. It points to the best possible course, whether or not the forecasts are accurate.

15

Decode: the term for the receiver translating a message into an idea or meaning.

Decomposition: the process of breaking down a project deliverable into smaller, more manageable components. In the Create WBS process, the results of decomposition are deliverables, whereas in the Define Activities process, project deliverables are further broken down into schedule activities.

Design of experiments: a statistical method for identifying which factors may influence specific variables of a product or process either under development or in production.

Direct costs: costs incurred directly by a project.

Effective listening: the receiver attentively watches the sender to observe physical gestures and facial expressions. In addition, the receiver contemplates responses, asks pertinent questions, repeats or summarizes what the sender has sent, and provides feedback.

Encode: the term for the sender translating an idea or meaning into a language for sending.

Enterprise environmental factors: external or internal factors that can influence a project's success. These factors include controllable factors such as the tools used in managing projects within the organization and uncontrollable factors that have to be considered by the project manager such as market conditions or corporate culture.

Expert judgment: judgment based on expertise appropriate to the activity. It may be provided by any group or person, either within the organization or external to it.

Expert power: power that results from an individual's knowledge, skills, and experience.

Fallback plan: a response plan that will be implemented if the primary response plan is ineffective.

Fast tracking: overlapping or performing in parallel project activities that would normally be done sequentially. Fast tracking may increase rework and project risk.

Feedback: affirming understanding and providing information.

Finish-to-finish: a logical relationship in which the predecessor must finish before the successor can finish.

Finish-to-start: a logical relationship in which the predecessor must finish before the successor can start. This is the default relationship for most software packages.

Fixed costs: nonrecurring costs that do not change if the number of units is increased.

Float: the amount of time that a scheduled activity can be delayed without delaying the end of the project. It is also called slack or total float. Float is calculated using a forward pass (to determine the early start and early finish dates of activities) and a backward pass (to determine the late start and late finish dates of activities). Float is calculated as the difference between the late finish date and the early finish date. The difference between the late start date and the early start date always produces the same value for float as the preceding computation.

Forward pass: used to determine the early start and early finish dates of activities.

Gantt chart: a bar chart that shows activities against time. Although the traditional early charts did not show task dependencies and relationships, modern charts often show dependencies and precedence relationships. These popular charts are useful for understanding project schedules and for determining the critical path, time requirements, resource assessments, and projected completion dates.

Good practice: a specific activity or application of a skill, tool, or technique that has been proven to contribute positively to the execution of a process.

Grade: the category or level of the characteristics of a product or service.

Hammock: summary activities used in a high-level project network diagram.

15

Heuristics: rules of thumb for accomplishing tasks. Heuristics are easy and intuitive ways to deal with uncertain situations; however, they tend to result in probability assessments that are biased.

Indirect costs: costs that are part of doing business and are shared among all ongoing projects.

Input: a tangible item internal or external to the project that is required by a process for the process to produce its output.

Issue: a risk event that has occurred.

Knowledge area: a collection of processes, inputs, tools, techniques, and outputs associated with a topical area. Knowledge areas are a subset of the overall project management body of knowledge that recognizes "good practices."

Lag: the amount of time a successor's start or finish is delayed from the predecessor's start or finish. In a finish-to-start example, activity A (the predecessor) must finish before activity B (the successor) can start. If a lag of three days is also defined, it means that B will be scheduled to start three days after A is scheduled to finish.

Lead (negative lag): the amount of time a successor's start or finish can occur before the predecessor's start or finish. In a finish-to-start example, activity A (the predecessor) must finish before activity B (the successor) can start. A lead of three days means that B can be scheduled to start three days before A is scheduled to finish.

Leadership: the ability to get an individual or group to work toward achieving an organization's objectives while accomplishing personal and group objectives at the same time.

Lean Six Sigma: a business improvement methodology that strives to eliminate non-value added activities and waste from processes and products.

Legitimate power: formal authority that an individual holds as a result of his or her position.

Letter contract: a written preliminary contract authorizing the seller to begin work immediately; it is often used for small value contracts.

Letter of intent: this is NOT a contract but simply a letter, without legal binding, that says the buyer intends to hire the seller.

Logical relationships: there are four logical relationships between a predecessor and a successor—finish-to-start, finish-to-finish, start-to-start, and start-to-finish.

Malcolm Baldrige: the national quality award given by the United States' National Institute of Standards and Technology. Established in 1987, the program recognizes quality in business and other sectors. It was inspired by Total Quality Management.

Management reserve: a dollar value, not included in the project budget, that is set aside for unplanned changes to project scope or time that are not currently anticipated.

Milestone chart: a bar chart that only shows the start or finish of major events or key external interfaces (e.g., a phase kickoff meeting or a deliverable); a milestone consumes NO resource and has NO duration; these charts are effective for presentations and can be incorporated into a summary Gantt chart.

Network diagram: a schematic display of project activities showing task relationships and dependencies; the PDM is useful for forcing the total integration of the project schedule, for simulations and "what-if" exercises, for highlighting critical activities and the critical path, and for determining the projected completion date.

Noise: anything that compromises the original meaning of a message.

Nonverbal communication: about 55% of all communication, based on what is commonly called body language.

Operation: ongoing work performed by people, constrained by resources, planned, executed, monitored, and controlled. Unlike a project, operations are repetitive; e.g., the work performed to carry out the day-to-day business of an organization is operational work.

Opportunities: risk events or conditions that are favorable to the project.

Opportunity costs: costs of choosing one alternative over another and giving up the potential benefits of the other alternative.

Organizational breakdown structure (OBS): different from a responsibility assignment matrix. The OBS is a type of organizational chart in which work package responsibility is related to the organizational unit responsible for performing that work. It may be viewed as a very detailed use of a RAM with work packages of the work breakdown structure (WBS) and organizational units as its two dimensions.

Organizational process assets: any formal or informal processes, plans, policies, procedures, guidelines, and on-going or historical project information such as lessons learned, measurement data, project files, and estimates versus actuals.

Organizational project management maturity model (OPM3®): focuses on the organization's knowledge, assessment, and improvement elements.

Output: a deliverable, result, or service generated by the application of various tools or techniques within a process.

Paralingual communication: optional vocal effects, the tone of voice that may help communicate meaning.

Penalty (coercive, punishment) power: power that results from an ability to take away something of value to another.

Percent complete: the amount of work completed on an activity or WBS component.

Performance domain: a broad category of duties and responsibilities that define a role. A performance domain expresses the actual actions of the project manager in that particular domain.

Phase: one of a collection of logically related project activities usually resulting in the completion of one or more major deliverables. A project phase is a component of a project life cycle.

15

Planning package: a component of the work breakdown structure that is below the control account to support known uncertainty in project deliverables. Planning packages will include information on a deliverable but without any of the details associated with schedule activities.

Point of total assumption: in a fixed price contract, the point above which the seller will assume responsibility for all costs; it generally occurs when the contract ceiling price has been exceeded.

Portfolio: a collection of programs, projects, and additional work managed together to facilitate the attainment of strategic business goals.

Power: the ability to influence people in order to achieve needed results.

Precision: a measure of exactness.

Predecessor: the activity that must happen first when defining dependencies between activities in a network.

Privity: the contractual relationship between the two parties of a contract. If party A contracts with party B, and party B subcontracts to party C, there is no privity between party A and party C.

Process: a collection of related actions performed to achieve a predefined desired outcome. The *PMBOK® Guide* defines a set of 47 project management processes, each with various inputs, tools, techniques, and outputs. Processes can have predecessor or successor processes, so outputs from one process can be inputs to other processes. Each process belongs to one and only one of the five process groups and one and only one of the ten knowledge areas.

Process group: a logical grouping of a number of the 47 project management processes. There are five process groups, and all are required to occur at least once for every project. The process groups are performed in the same sequence each time: initiating, planning, executing, more planning and executing as required, and ending with closing. The monitoring and controlling process group is performed throughout the life of the project. Process groups can be repeated for each phase of the project life cycle. Process groups are not phases. Process groups are independent of the application area or the life cycle utilized by the project.

Process quality: specific to the type of product or service being produced and the customer expectations, the level of process quality will vary. Organizations strive to have efficient and effective processes in support of the product quality expected. For example, the processes associated with building a low-quality, low-cost automobile can be just as efficient, if not more so, than the processes associated with building a high-quality, high-cost automobile.

Product life cycle: the collection of stages that make up the life of a product. These stages are typically introduction, growth, maturity, and retirement.

Product quality: specific to the type of product produced and the customer requirements, product quality measures the extent to which the end product(s) of a project meets the specified requirements. Product quality can be expressed in terms that include, but are not limited to, performance, grade, durability, support of existing processes, defects, and errors.

Program: a group of related projects managed in a coordinated way; e.g., the design and creation of the prototype for a new airplane is a project, while manufacturing 99 more airplanes of the same model is a program.

Progressive elaboration: the iterative process of increasing the level of detail in a project management plan as greater amounts of information and more accurate estimates become available.

Project: work performed by people, constrained by resources, planned, executed, monitored, and controlled. It has definite beginning and end points and creates a unique outcome that may be a product, service, or result.

Project life cycle: the name given to the collection of various phases that make up a project. These phases make the project easier to control and integrate. The result of each phase is one or more deliverables that are utilized in the next few phases. The work of each phase is accomplished through the iterative application of the initiating, planning, executing, monitoring and controlling, and closing process groups.

Project management: the ability to meet project requirements by using various knowledge, skills, tools, and techniques to accomplish project work. Project work is completed through the iterative application of initiating, planning, executing, monitoring and controlling, and closing process groups. Project management is challenged by competing and changing demands for scope (customer needs, expectations, and requirements), resources (people, time, and cost), risks (known and unknown), and quality (of the project and product).

Project management information system: the collection of tools, methodologies, techniques, standards, and resources used to manage a project. These may be formal systems and strategies determined by the organization or informal methods utilized by project managers.

Project management methodology: any structured approach used to guide the project team through the project life cycle. This methodology may utilize forms, templates, and procedures standard to the organization.

Project network schedule calculations: there are three types of project network schedule calculations—a forward pass, a backward pass, and float. A forward pass yields early start and early finish dates, a backward pass yields late start and late finish dates, and these values are used to calculate total float.

Project quality: typically defined within the project charter, project quality is usually expressed in terms of meeting stated schedule, cost, and scope objectives. Project quality can also be addressed in terms of meeting business objectives that have been specified in the charter. Solving the business problems for which the project was initiated is a measure of the quality for the project.

15

Quality: the degree to which a set of inherent characteristics satisfies the stated or implied needs of the customer. To measure quality successfully, it is necessary to turn implied needs into stated needs via project scope management.

Quality objective: a statement of desired results to be achieved within a specified time frame.

Quality policy: a statement of principles for what the organization defines as quality.

Referent (charisma) power: power that results from a project manager's personal characteristics.

Requirements traceability matrix: a matrix for recording each requirement and tracking its attributes and changes throughout the project life cycle to provide a structure for changes to product scope. Projects are undertaken to produce a product, service, or result that meets the requirements of the sponsor, customer, and other stakeholders. These requirements are collected and refined through interviews, focus groups, surveys, and other techniques. Requirements may also be changed through the project's configuration management activities.

Residual risk: when implementing a risk response plan, the risk that cannot be eliminated.

Resource calendar: a calendar that documents the time periods in which project team members can work on a project.

Resource optimization techniques: techniques that are used to adjust the start and finish dates of activities that adjust planned resource use to be equal to or less than resource availability.

Responsibility assignment matrix (RAM): a structure that relates project roles and responsibilities to the project scope definition.

Reward power: power that results from an ability to give something of value to another.

Risk: an uncertain event or condition that could have a positive or negative impact on a project's objectives.

Risk appetite: the degree of uncertainty an entity is willing to take on in anticipation of a reward.

Risk threshold: the measures, along the level of uncertainty or the level of impact, at which a stakeholder may have a specific interest. Risk will be tolerated under the threshold and not tolerated over the threshold.

Risk tolerance: the degree, amount, or volume of risk that an organization or individual will withstand.

Rolling wave planning: a progressive elaboration technique that addresses uncertainty in detailing all future work for a project. Near-term work is planned to an appropriate level of detail; however, longer term deliverables are identified at a high level and decomposed as the project progresses.

Schedule activity: an element of work performed during the course of a project. It is a smaller unit of work than a work package and the result of decomposition in the Define Activities process of project time management. Schedule activities can be further subdivided into tasks.

Scheduling charts: there are four types of scheduling charts— the Gantt chart, the milestone chart, the network diagram, and the time-scaled network diagram.

Scope baseline: the approved detailed project scope statement along with the WBS and WBS dictionary.

Scope creep: the uncontrolled expansion of a product or project scope without adjustments to time, cost, and resources.

Secondary risk: when implementing a risk response, a new risk that is introduced as a result of the response.

Seller: the bidder, contractor, source, subcontractor, vendor, or supplier who will provide the goods and services to the buyer. The seller generally manages the work as a project, utilizing all processes and knowledge areas of project management.

Single source: selecting a seller without competition. This may be appropriate if there is an emergency or prior business relationship.

Six Sigma: an organized process that utilizes quality management for problem resolution and process improvement. It seeks to identify and remove the causes of defects.

Sole source: selecting a seller because it is the only provider of a needed product or service.

Stakeholder: an individual, group, or organization who may affect, be affected by, or perceive itself to be affected by a decision, activity, or outcome of a project.

Stakeholder register: a project document that includes the identification, assessment, and classification of project stakeholders.

Standard: a document that describes rules, guidelines, methods, processes, and practices that can be used repeatedly to enhance the chances of success.

Standard deviation: the measurement of the variability of the quantity measured, such as time or costs, from the average.

Start-to-finish: a logical relationship in which the predecessor must start before the successor can finish; this is the least used and some software packages do not even allow it.

Start-to-start: a logical relationship in which successor can start as soon as the predecessor starts.

Statistical sampling: involves choosing part of a population of interest for inspection.

Statistical terms: the primary statistical terms are the project mean, variance, and standard deviation.

Subproject: a component of a project. Subprojects can be contracted out to an external enterprise or to another functional unit.

Successor: the activity that happens second or subsequent to a previous activity when defining dependencies between activities in a network.

Sunk costs: money already spent; there is no more control over these costs. Since these are expended costs they should not be included when determining alternative courses of action.

Tailoring: the act of carefully selecting processes and related inputs and outputs contained within the *PMBOK® Guide* to determine a subset of specific processes that will be included within a project's overall management approach.

Team building: the process of getting a diverse group of individuals to work together effectively. Its purpose is to keep team members focused on the project goals and objectives and to understand their roles in the big picture.

Technique: a defined systematic series of steps applied by one or more individuals using one or more tools to achieve a product or result or to deliver a service.

Threat: risk events or conditions that are unfavorable to a project.

Time-scaled network diagram: a combination of a network diagram and a bar chart that shows project logic, activity durations, and schedule information.

Tool: a tangible item such as a checklist or template used in performing an activity to produce a product or result.

Transmit message: the term for using a communication method to deliver a message.

Triangular distribution or **three-point estimating**: takes the average of three estimated durations—the optimistic value, the most likely value, and the pessimistic value. By using the average of three values rather than a single estimate, a more accurate duration estimate for the activity is obtained.

Variable costs: costs that increase directly with the size or number of units.

Virtual teams: groups of people with shared objectives who fulfill their roles with little or no time spent meeting face to face.

15

Warranties: assurance that the products are fit for use or the customer receives compensation. Warranties could cover downtime and maintenance costs.

WBS dictionary: houses the details associated with the work packages and control accounts. The level of detail needed will be defined by the project team.

Weighted three-point estimating or **beta/PERT**: the program evaluation and review technique (PERT) uses the three estimated durations of three-point estimating but weighs the most likely estimate by a factor of four. This weighted average places more emphasis on the most likely outcome in calculating the duration of an activity. Therefore, it produces a curve that is skewed to one side when possible durations are plotted against their probability of occurrence.

What-if scenario analysis: the process of evaluating scenarios in order to predict their effect on project objectives.

Work breakdown structure (WBS): a framework for defining project work into smaller, more manageable pieces, it defines the total scope of the project using descending levels of detail.

Work package: the lowest level of a WBS; cost estimates are made at this level.

Workarounds: unplanned responses to risks that were previously unidentified or accepted.

15

APPENDIX D

BIBLIOGRAPHY

This study guide and the *PMBOK® Guide* provide enough material for most experienced project managers to pass the exam. However, if you feel that you need additional materials, here are some books that we have found useful and that we reference in the study guide.

Brake, Terence, Danielle Walker, and Thomas Walker. *Doing Business Internationally: The Guide to Cross-cultural Success.* New York: McGraw-Hill, 1995.

Cleland, David, Karen M. Bursic, Richard Puerzer, and A. Yaroslav Vlasak, eds. *Project Management Casebook.* Newtown Square, PA: Project Management Institute, 1998.

Clemen, Robert T. *Making Hard Decisions: An Introduction to Decision Analysis*, Second Edition. Pacific Grove, CA: Duxbury Press, 1996.

Ferraro, Gary P. *The Cultural Dimension of International Business*, Fourth Edition. Upper Saddle River, NJ: Prentice-Hall, 2002.

Fleming, Quentin W. and Joel M. Koppelman. *Earned Value Project Management*, Second Edition. Newtown Square, PA: Project Management Institute, 1996.

Garner, Bryan A., ed. *Black's Law Dictionary*, Eighth Edition. New York: Thomson West, 2004.

Ireland, Lewis R. *Quality Management for Projects and Programs.* Newtown Square, PA: Project Management Institute, 1991.

Kerzner, Harold. *Project Management: A Systems Approach to Planning, Scheduling, and Controlling*, Eighth Edition. New York: John Wiley & Sons, 2003.

15

Lewis, James P. *Fundamentals of Project Management*. New York: American Management Association, 1997.

----------. *Project Planning, Scheduling and Control*. Chicago: Probus Publishing, 1991.

Meredith, Jack R. and Samuel J. Mantel, Jr. *Project Management: A Managerial Approach*, Fourth Edition. New York: John Wiley & Sons, 2000.

Project Management Institute. *A Guide to the Project Management Body of Knowledge*, Fifth Edition (*PMBOK® Guide*). Newtown Square, PA: Project Management Institute, 2012.

----------. *Code of Ethics and Professional Conduct*.

----------. *PMI Lexicon of Project Management Terms*.

----------. *PMI PMP Exam Content Outline—June 2015*.

----------. *Practice Standard for Estimating and Practice Standard for Earned Value Management*, Second Edition.

----------. *Principles of Project Management*. Newtown Square, PA: Project Management Institute, 1997.

----------. *Project Management Experience and Knowledge Self-Assessment Manual*. Newtown Square, PA: Project Management Institute, 2000.

----------. *Project Management Professional (PMP) Examination Specification*. Newtown Square, PA: Project Management Institute, 2005.

Rosen, Robert, Patricia Digh, Marshall Singer, and Carl Phillips. *Global Literacies: Lessons on Business Leadership and National Cultures*. New York: Simon & Schuster, 2000.

Rosenau, Milton D., Jr. *Successful Project Management, A Step-by-Step Approach with Practical Examples*, Third Edition. New York: John Wiley & Sons, 1998.

Trompenaars, Fons and Charles Hampden-Turner. *Riding the Waves of Culture: Understanding Diversity in Global Business*, Second Edition. New York: McGraw-Hill, 1998.

Verma, Vijay K. *Human Resource Skills for the Project Manager*. Newtown Square, PA: Project Management Institute, 1996.

----------. *Managing the Project Team*. Newtown Square, PA: Project Management Institute, 1995.

----------. *Organizing Projects for Success*. Newtown Square, PA: Project Management Institute, 1995.

Wideman, R. Max, ed. *A Framework for Project and Program Management Integration*. Newtown Square, PA: Project Management Institute, 1991.

----------. *Project and Program Risk Management: Guide to Managing Project Risks and Opportunities*. Newtown Square, PA: Project Management Institute, 1992.

Wysocki, Robert K., Robert Beck Jr., and David B. Crane. *Effective Project Management: How to Plan, Manage, and Deliver a Project on Time and Within Budget*. New York: John Wiley & Sons, 2000.

15

16

16

16

16